MATH ON TRIAL

MATH ON TRIAL

How Numbers Get Used and Abused in the Courtroom

Leila Schneps and Coralie Colmez

BASIC BOOKS

A Member of the Perseus Books Group • *New York*

Books published by Basic Books are available at special discounts for bulk purchases in the United States by corporations, institutions, and other organizations. For more information, please contact the Special Markets Department at the Perseus Books Group, 2300 Chestnut Street, Suite 200, Philadelphia, PA 19103, or call (800) 810-4145, ext. 5000, or e-mail special.markets@perseusbooks.com.

Designed by Jeff Williams

Library of Congress Cataloging-in-Publication Data
Schneps, Leila.
   Math on trial : how numbers get used and abused in the courtroom / Leila Schneps and Coralie Colmez.
      pages cm
   Includes bibliographical references and index.
   ISBN 978-0-465-03292-1 (hardcover : alk. paper) — ISBN 978-0-465-03794-0 (e-book)
   1. Forensic statistics. I. Colmez, Coralie, 1988– II. Title.

K2290.S73S36 2013
345'.07—dc23

                                                                      2012040624

10 9 8 7 6 5 4 3 2 1

We dedicate this book to all those who have suffered from miscarriages of justice, and to all victims of crimes whose perpetrators went unpunished, due to the misuse or misunderstanding of mathematics in the legal process.

CONTENTS

INTRODUCTION

Everywhere we turn we are assailed by numbers. Advertisements, news, price reductions, medical information, weather forecasts, investment, risk assessment: all this and more is communicated to us through probabilities and statistics. But the problem is that these figures are not always used to convey information. As often as not, they are used to give us spin: to influence, frighten, and mislead us with the cool authority of numbers and formulas.

Now, you might think this a trivial matter. You may be one of those people who skip past the numbers in the articles you read, who pay no attention to the declarations of sensational increase or decrease in whatever drama is playing out on the front page, whether it be global warming, shark infestations, or illiteracy. At worst, you think, people are mildly misinformed. But as we show in *Math on Trial*, the misuse of mathematics can be deadly. The same mathematical tricks that mislead the public about market trends and risk and social problems have sent innocent people to prison. Being wrong about the price of oil is one thing; being denied justice due to miscalculation is quite another.

Despite their ubiquity, however, most of these fallacies are easy to spot. The fact is that anyone can make a decent assessment of mathematical statements that appear in popular publications, on common products, and in everyday activities from investment to DNA analysis. Anyone can acquire the simple reflexes to cut through the fog of mathematical deception. All it takes is a little practice to recognize what's going on. It turns out there isn't much variation in these numerical sleights of hand, but public ignorance allows them to permeate every area of our lives. We have chosen the examples in this book because while illustrating the pitfalls that everyone should be aware of, they also show that the misuse of mathematics is not merely an academic issue that we can easily ignore.

We need to know when we are being misled. We need to be able to distinguish whether the numbers brandished in our faces are legitimately providing information or being misused for dangerous ends. We need to go beyond the abstraction of theory and see the plain truth for ourselves.

Mathematics has made but few appearances in criminal trials throughout history. When it has been used, it has been for purposes of identification, to calculate the probability that a given identification is correct. These same calculations occur in a thousand other domains of public and private life, and one might wonder why we have chosen to focus here on its relatively rare use in trials. We believe it is worth collecting and examining these cases for the simple reason that many of the common mathematical fallacies that pervade the public sphere are perfectly represented by these trials. Thus, they serve as ideal illustrations of these errors and of the drastic consequences that faulty reasoning has on real lives.

The cases we present in this book cover a broad range of mathematics used in the courtroom, from the simplest handwriting analysis at the end of the nineteenth century to probabilities used in DNA identification today. These cases are not ordered chronologically, but according to the complexity of the probability concepts in question. We discuss cases in which mathematics was presented at trial to justify conviction, and others in which it was employed to convince the public that conviction was erroneous.

In spite of mathematics' disastrous record of causing judicial error, the main conclusion of our analysis is not that probability is a useless cog in the judicial machine. Rather, we found that the injustices perpetrated in the name of probability arise from the misuse of mathematical principles, not from any inherent inapplicability of mathematics to justice. We believe that mathematics can be useful in fundamental ways, and indeed that the future of criminal justice will necessarily contain an element of mathematical analysis, given the prevalence of DNA evidence in trials today. But to reach that goal there must be some certainty that mathematical errors will be excluded from trials, and the first step in this direction is to identify the most important errors that have actually occurred.

In this book we share the dramas of people who saw their lives ripped apart by simple mathematical errors—wrong calculations, or calculations that were not made or not understood—grave injustices that were committed or only narrowly avoided. We hope that these incredible true stories will show that mathematics can really be a matter of life and death.

MATH ERROR NUMBER 1 »
MULTIPLYING NON-INDEPENDENT PROBABILITIES

MOST PEOPLE KNOW that to measure the probability that several events will occur, the separate probabilities of each event should be multiplied together. For instance, if you are pregnant with a single child, there is a 1 out of 2 chance you will give birth to a girl. Thus if you have two children at different times, the probability of having 2 girls is ½ squared, which is ¼, or 1 chance out of 4.

We do this type of calculation all the time, almost without thinking. But there's a caveat: this multiplication is correct only if the events you're comparing are totally independent from each other, like having separate pregnancies. If they are not independent, the situation changes. Suppose, for example, that you happen to know from an ultrasound that you are pregnant with identical twins. Now the birth of your two children does not constitute two independent events, and of course it would be wrong to say that the probability of your having two girls is ¼; it is in fact ½, because the two babies share the same genes, so they will necessarily be of the same sex; thus they can only be either two girls or two boys.

If you multiply the probabilities of events that are not independent of each other, you will get a significantly smaller probability than is accurate. But it's easy to fall into the trap of assuming that a set of separate events occurred or will occur independently of one another. Some events may seem independent but have a single underlying cause. For example, a card player may go on a winning streak that defies all odds—but the reason could be that he's cheating.

It's risky to assume that events are independent when all the data is not in. Yet it has been done, even by highly respected people, in courts of law. And sometimes it has resulted in disaster.

The Case of Sally Clark:
Motherhood Under Attack

Steve and Sally Clark were a loving couple of bright, ambitious young lawyers. Both worked demanding jobs in London, but eventually they bought themselves a little house called Hope Cottage, well away from the bustle of the city, and decided to raise a family. On September 22, 1996, Sally gave birth to a son, Christopher. She decided to stop working for a few months and stay home with her child.

From the beginning the baby appeared fragile and delicate, with the face of an angel. He was extremely quiet, slept a great deal, and almost never cried. In early December he developed what seemed to be sniffles and a bad cold, but the doctor told Sally not to worry. Everything seemed normal enough until December 13, when she went down to the kitchen for ten minutes to prepare herself a drink, and returned to the bedroom to find the baby gray-faced in his basket. She called for an ambulance and the baby was rushed to the hospital, but sadly he could not be saved. An autopsy indicated that he had been suffering from an infection of the lungs.

After Christopher's death Sally returned to work, but although she functioned adequately, she underwent a period of grieving, depression, and despair, occasionally drinking heavily. A new pregnancy helped her snap out of it, and she underwent therapy to bolster a complete renunciation of alcohol. Healthy baby Harry was born on November 29, 1997.

Like all younger siblings of babies who have died in England, this second baby was closely monitored under the program known as Care of Next

Infants (CONI). Steve and Sally were taught the basic gestures of resuscitation, and Harry was given an apnea alarm to wear permanently, which was supposed to start ringing if he stopped breathing. As a matter of fact, the alarm went off quite frequently, but the health visitors and nurses who stopped by the house both regularly and for random checkups found nothing wrong with the baby, so everyone assumed that the apnea alarm was malfunctioning. Little Harry appeared strong and hearty, was noisy and active, cried loudly, and demanded frequent feedings. Sally devoted herself to him and kept a close eye on his health, filling out the many charts required by the CONI program and keeping him well away from any chance of infection by contact with other sick people. As Steve was in a cast with a torn Achilles tendon, the Clarks hired daily help during the first weeks of Harry's life to give Sally a hand with all the housework. On January 26, 1998, Sally took Harry to the community health center for his standard vaccinations.

After the vaccinations he was much quieter than usual, appearing lethargic and pale as Sally wheeled him home. Five hours later Steve was trying to amuse the baby and play with him, but Harry didn't seem interested, so Steve put him down in his bouncy chair and went to the kitchen. Not five minutes later he heard Sally calling him desperately. Little Harry had gone limp and white; his head was falling forward. Steve rushed back to the bedroom, laid the baby on the floor, and tried to resuscitate him, first gently and then with increasing strength, while Sally called for help. The ambulance arrived and rushed the family to the hospital. But for the second time, doctors were unable to save the life of the couple's baby.

This time the autopsy gave surprising, and seemingly contradictory, evidence. The pathologist, Dr. Williams, claimed that he could see retinal hemorrhage in Harry's eyes, a frequent sign of smothering, and could feel a broken rib, though whether recent or old he could not say; it did not show up on X-ray. Harry also had large amounts of bacteria in his nose, throat, lungs, and stomach, but no notice was taken of this. The pathologist believed there was sufficient evidence of abuse to warrant a complete investigation.

Steve and Sally Clark were arrested for the murder of their two children. After intensive interrogation, during which they answered all questions freely and openly and did not ask for a lawyer, they were released on bail while the investigation proceeded.

Steve and Sally Clark

They returned home, minus their passports and with the obligation of registering regularly at the police station, and tried to pick up the remains of their shattered lives. But to their horror, as the investigation continued and they were repeatedly called in for questioning, they realized that their desperate need to understand the medical causes for the death of their sons was gradually being overshadowed by the new, urgent necessity of defending themselves from the accusations of severe child abuse being leveled against them by the police. They realized that they had no adequate defense against such an accusation—there is no actual proof that a dead baby has *not* been smothered! They could hardly believe that the investigation would result in a trial, yet on the advice of their friends they eventually went to a criminal lawyer. Solicitor Mike Mackey agreed to take their case and help them, come what may.

Two important events followed: a third little boy, born a year to the day after Harry's birth, and a formal charge of double murder against Sally.

Steve, exonerated from any wrongdoing and not charged, was helpless to prevent the destruction of his family and the persecution of his wife. The new baby was placed in foster care, and a date was set for Sally's trial for the murder of her sons.

Sally's trial took place at Chester Crown Court, before a judge and jury. She was defended by brilliant lawyers who put their finger squarely on each and every one of the contradictions in the massive and complex medical testimony, obliging the medical experts to contradict each other, and pointing out a series of errors of interpretation in Harry's autopsy. Most of the prosecution's experts were forced to admit that the deaths of the babies could not be definitively attributed to shaking, smothering, or other abuse, and Sally's behavior as a mother was vouched for by many witnesses, such as the nanny who had helped her with Harry, and the health care professionals who had kept him under regular observation for the CONI program. Listening to them, Sally felt certain that her innocence could only be obvious to the jury. It was this certainty, this faith in the justice system, that gave her the strength to sit through grueling hours of description of the autopsies of her sons, in which every sign of possible violence was discussed in gruesome but unavoidable detail. During those hours of testimony from the prosecution's medical experts, Sally was forced to listen to the hateful picture that was being painted for the jury of the person she was supposed to be—obsessively tidy, professionally ambitious, a control freak, unfit to be a mother—and the actions she was accused of having committed. Not only she, but also the spectators at the trial were horrified by a system that imposes such torment on parents who have lost their children. Was it really necessary for Steve Clark, gagging on the witness stand, to be shown photographs of his dead babies' medically dismembered little bodies?

Everything the medical experts were saying seemed wrong to Sally—drastically, obviously, cruelly, horribly, and offensively wrong. Until the renowned pediatrician Sir Roy Meadow took the stand.

Charming and avuncular, Meadow appeared filled with sympathy for the plight of the accused, pronouncing words of condemnation with a seeming reluctance that made his allegations all the more effective. He exuded competence, experience, ability, and kindness. Listening to his testimony, Sally was struck dumb. "If I didn't know I was innocent," she later said, "listening to him I would have believed myself guilty."

Dr. Roy Meadow, pediatrician

And up on the witness stand, Meadow spoke the words that swung the balance of justice irrevocably against Sally.

IN ORDER to understand what Roy Meadow was doing when he told the judge and jury his own opinions about Sally Clark and the death of her babies, and why his statements carried such weight, it is important to know who he was, where he was coming from, and where his sphere of competency lay. A specialist in child abuse, he had studied under legendary child psychoanalyst Anna Freud, and was greatly influenced by her teaching. "A child needs mothering—not a mother," he used to quote her as saying, though it is not absolutely certain that Freud ever really pronounced this sentence; perhaps those words were merely Meadow's own interpretation of what she taught. In any case, they seem to have left their mark.

While Roy Meadow began as a pediatrician, working first as a general practitioner, and later at Guy's Hospital, the Hospital for Sick Children in London, and the Royal Alexandra Hospital in Brighton, his main interest changed to child abuse as his career advanced, and he focused his attention on the detection, analysis, and proof of the misdeeds and cruelties of mothers. It was in 1977, while working as a senior lecturer and consultant pediatrician at Leeds University, that Meadow came up with the idea that eventually led him to fame. This was the discovery—or the invention—of a new malady, which he baptized *Munchausen Syndrome by Proxy*.

Munchausen Syndrome is the name given by Dr. Richard Asher in 1951 to a psychological condition by which a person who is actually in perfectly good health claims to suffer from all kinds of symptoms of illness that he imagines to be truly present, or sometimes even purposely brings on through acts of self-injury. The name is a reference to the yarns told by the eighteenth-century German soldier and nobleman Baron Munchausen, who astounded his listeners upon his return from the wars by describing flights on cannonballs, trips to the moon, and impressive feats of marksmanship such as shooting fifty brace of ducks with a single bullet. There is actually not a lot of resemblance between Munchausen's tales and those told by sufferers from Munchausen Syndrome—except, perhaps, for their tallness.

Psychological analysis has determined that Munchausen Syndrome arises from an intense need for sympathy, care, and attention from a competent and protective figure, a role that is ideally played by a doctor. Exactly the kind of medical test or procedure that most people would prefer to avoid—blood tests, biopsies, colonoscopies—is reassuring and consoling to sufferers of Munchausen Syndrome, and they tend to seek such procedures repeatedly and unnecessarily.

What Roy Meadow noticed in his seminal 1977 paper was that some people display a variant form of Munchausen Syndrome, seeking constant medical attention not for themselves, but for another person, a "proxy." These people constantly go to doctors and describe symptoms in their proxy that are either nonexistent or artificially induced. Obviously the proxy must be someone unable to explain the true state of affairs; for this reason, proxies tend to be helpless invalids or children.

This was the mental condition that Meadow dubbed Munchausen Syndrome by Proxy (MSbP). He published his discovery in the medical journal *The Lancet,* and the title of the article reveals that his interest in the syndrome itself was inspired by a profound concern with the frightening realities of child abuse. The article, "Munchausen Syndrome by Proxy: the hinterland of child abuse," described two cases in which he had taken a particular interest. In one of them, a mother persistently altered her six-year-old daughter's perfectly healthy urine samples, leading doctors to perform an endless stream of invasive medical examinations on the child and subject her to long-term medicinal treatments ranging from antibiotics to chemotherapy. The deception stopped only when the daughter was admitted to a hospital and kept for two or three days in the absence of her mother, who previously had rarely left her side. The child's healthy samples during the period of her mother's absence, and her instant relapse the moment her mother returned, finally led caregivers to the truth. In the second case, a seemingly devoted mother brought her toddler to the hospital at least once a month with attacks of illness that were diagnosed as salt poisoning. When the child was kept in the hospital, he became healthy, and when his mother

visited, he relapsed. The hospital contacted social services to organize surveillance and placement for the child, but before the discussions could lead
to a concrete result, the little boy was brought in with an attack so severe
that he died.

If the hospital workers took as long as they did to detect what was going
on, Roy Meadow explained, it was partly because both mothers seemed to
be agreeable, intelligent women and loving and tender parents (albeit with a
history of hysterical behavior, had anyone only thought to check). No one
suspected them, because no one had the habit of suspecting mothers. Roy
Meadow stressed the fact that mothers *must* be suspected. "We may teach,
and I believe should teach, that mothers are always right," he wrote, "but at
the same time we must recognize that when mothers are wrong they can be
terribly wrong."

For ten or twelve years after Meadow wrote his article, Munchausen Syndrome by Proxy received little or no attention, either in the profession or from
the public. And then suddenly, he was given the opportunity to put his theory
into practice, because of a grisly and terrible murder case that finally allowed
the whole idea to spring forth and capture the attention of a country.

IN FEBRUARY 1991, a young nurse called Beverley Allitt was engaged in Children's Ward 4 of the severely understaffed Grantham and Kesteven Hospital
in Lincolnshire. Although she appeared kind and competent, she inexplica-
bly went on a killing spree that, in the space of barely two months, took the
lives of four tiny children and severely injured one more.

Recalling what it was like to work alongside Beverley Allitt for those two
months, nurse and coworker Mary Reet expressed an intuitive feel for Allitt's
psychological motivation. "Part of the kick she would've had was that when
those babies were brought back to life, she was there, and she was the savior,"
she later wrote. It wasn't the babies' deaths that Allitt wanted: it was attention.
And this was exactly how Roy Meadow presented it when he testified as a
medical expert at her trial. He showed how she displayed all the symptoms
of both Munchausen Syndrome and MSbP, and explained that Beverley's
coldness in the face of the death of her victims was typical; people with
MSbP are not able to grasp the harm they are inflicting; they are closed off
to it. Roy Meadow stated that he did not believe Beverley Allitt could be

cured. She would always be a danger to others. Allitt was convicted and given thirteen life sentences.

Roy Meadow's diagnosis made a lot of sense. On top of that, the visibility of the case, the terrible and shocking nature of the crimes, and his role as expert witness at the trial conferred upon him not just fame, but a great deal of influence and power as well.

From the moment Beverley Allitt was condemned to life in prison, Roy Meadow's theory of Munchausen Syndrome by Proxy took a tremendous leap into both the public and the medical consciousness. It is perhaps not fully realized how rapidly the notion took hold and the number of diagnosed cases grew. MSbP became a byword in social work, where it was cited as a reason to interfere in the lives of innumerable families. Thousands of children were removed from their parents, and the practice soon spread from Britain to the United States, then to Australia, New Zealand, Germany, and Canada, and as far as Nigeria and India, where it was very popular.

In the name of the new diagnosis, horrific mistakes were made. One example was the "legal kidnapping" of little Philip P. in the autumn of 1996. Philip was a child less than one year old whose mother, Julie P., had made countless trips to hospitals in the state of Tennessee, where she lived, seeking treatment for her baby's severe birth defects and chronic gastrointestinal troubles. Seeing the child's lengthy medical history, doctors concluded that Julie must suffer from MSbP, and shortly after the baby's arrival at the hospital they contacted the Department of Children's Services to have custody of the baby removed from his parents and given to the hospital. The family was kept at a distance and the nurses were even instructed to give them no medical details over the phone. Unfortunately, as it happened, the baby was really very ill; separation from the mother had no curative effects, and exactly one month later he died, alone and far from his parents. It was one of many cases in which the notion of MSbP was abused.

ALLITT WAS convicted in 1993, and by 1996 some doctors were already beginning to sound a warning bell, even as allegations of MSbP continued to increase at a terrifying pace. Dr. C. J. Morley published an article titled "Practical concerns about the diagnosis of Munchausen syndrome by proxy," in which he warned that after the condemnation of Beverley Allitt, the diagnosis had become "charged with emotion" and that "those who are accused are tarnished with her reputation." In the article, he discussed the so-called symptoms of MSbP one by one, showing that each of them might arise for a perfectly legitimate reason. He even warned against a diagnosis of MSbP in the instance where a child separated from his mother is cured of his illness, as there can be any number of reasons for such an event, including the natural recovery from many infantile illnesses that tend to occur around the age of one year.

In their 1995 article "Is Munchausen syndrome by proxy really a syndrome?" G. Fisher and I. Mitchell also analyzed the weaknesses of the diagnosis, ending by suggesting it be dropped altogether: "It is recommended that pediatricians abandon making diagnoses of Munchausen Syndrome by Proxy and instead diagnose the specific fabricated or induced medical illness(es) or condition(s) they encounter."

But such calls for restraint were not heeded. Allegations of MSbP continued apace, and in fact a new aspect appeared and soon came to occupy a central position in allegations of child abuse: the role of Munchausen in unexplained deaths of babies, often referred to as cot death, crib death, or SIDS—Sudden Infant Death Syndrome.

BABIES HAVE always been fragile beings. The death rate of children under the age of one year old in the nineteenth century was startlingly high—as many as one hundred per thousand babies in the upper social classes and three hundred per thousand among the poor. Even in the early twentieth century the numbers remained significant. Only after the Second World War did doctors and hospitals begin to make tremendous strides in infant care, and rates began to decrease.

Yet even today, a small number of seemingly healthy very young babies die suddenly from unexplained causes. The phenomenon of crib death was not treated as a medical phenomenon in and of itself until 1963, when a first conference on the subject was organized in Seattle, Washington. The official term "Sudden Infant Death Syndrome" (SIDS) was adopted at a second conference in 1969. Obviously, the question of what proportion of SIDS deaths might be attributed to child abuse or outright murder was raised at both conferences, but there was simply no information available to draw any real conclusions as to the answers. A suggestion that

SIDS was connected to apnea (babies stopping breathing for no reason) led to the adoption of huge numbers of apnea alarms being installed. These detectors are attached by small sensors to the baby's body and are set to go off loudly if no breathing motion is detected. In the end, however, their widespread use served mainly to show that apnea is not the only or major cause of SIDS; there were too many cases where the alarm never went off. And studies showed that, like other social phenomena, SIDS is correlated with such factors as family background, poverty, mental illness, smoking, or drug use.

Improvements in baby care methods led to a significant drop in SIDS throughout the 1990s, particularly in families considered "low risk," meaning stable families with good incomes and good physical and mental health. The improvement spurred further study, and during the early 1990s there occurred a kind of medical SIDS frenzy, with doctors publishing research based on as few as two cases, rising to stellar heights in their careers, and, even worse, encouraging parents who might be prone to Munchausen-type behavior to give way to it completely by calling for them to bring their babies repeatedly to the hospital for examination and care, exactly the kind of treatment that MSbP patients thrive on.

Then the bubble burst.

A team of suspicious law enforcement officials demanded and obtained the exhumation and autopsy of three siblings from New York State who had all supposedly died of SIDS. A detailed medical and legal investigation eventually proved that in fact they had all been suffocated by their father. The same team then investigated a family in which two infants who had died of SIDS had been the subject of a highly respected medical publication on SIDS and apnea. They discovered that three older siblings of these two babies had also died. Their mother eventually confessed to all the murders. This event and other similar ones were the flash points that drew together two strands that had been unconnected until then: SIDS (crib death) and Munchausen Syndrome by Proxy.

Until the mid-1990s, Munchausen Syndrome by Proxy had been studied in parents or caregivers who harmed children in order to gain attention and care. The children sometimes died, but this had not appeared as the goal of the caregivers' actions. But then came the first diagnosis of Munchausen Syndrome by Proxy in a case of repeated SIDS.

ROY MEADOW'S intention, at first, was to join the swelling ranks of doctors concerned with finding causes and preventive remedies for SIDS, and to find features that could distinguish between natural SIDS and the death of a child caused by suffocation or other mistreatment. Since it is nearly impossible to detect signs of suffocation—the parents' desperate vigorous attempts at resuscitation may cause the same slight bruises or cracked ribs as intentional abuse—the possibilities for detecting the difference seemed slim. Yet they were of paramount importance, both for those babies who suffered "near-miss" crib death and for the siblings of those whose abuse had gone undetected. Like many other doctors involved in the movement, Dr. Meadow wanted to find some signs that could help him tell one from the other. He devoted himself to this subject in the 1997 book he edited, *The ABC of Child Abuse,* and in a study titled "Unnatural Sudden Infant Death," in which he surveyed eighty-one cases of sudden infant death, collected over an eighteen-year period, in families where the parents had actually been convicted of murder. He attempted to outline some general types of scenario to distinguish SIDS from murder. Unfortunately, the problem proved difficult; in half of the cases, autopsies showed physical signs of suffocation, but in the other half there were none.

It is a fact that some proportion of SIDS is unquestionably caused by parental abuse, but no one knows what that proportion is, and it is practically impossible to distinguish those cases with certainty. Until the 1990s the attitude of pediatricians toward this problem was to "let ill alone lest worse befall," in the words of Dr. John Davies, who worried that innocent parents would be accused, families broken for no reason, and siblings sent into foster care.

But Roy Meadow was convinced that there were far more parents murdering their babies than anyone had ever realized—or at least admitted. He came to believe that Munchausen Syndrome by Proxy, hitherto regarded as a phenomenon leading mothers to harm their children, was actually claiming the lives of a much larger number of tiny victims than anybody realized. From that point on, his book, his study, and all the rest of his work became focused on developing a newly hawkish "interventionist strategy," which meant making sure that mothers whose children died would be accused of having killed them if no other medical cause could be found. And his reputation as a specialist of the MSbP phenomenon lent tremendous weight to his words.

Partly because of that reputation and partly because of his vocal attitude toward child abuse, Roy Meadow's career skyrocketed. He became president of the British Paediatric Association in 1994 and president of the Royal College of Paediatrics and Child Health in 1996. In the New Year honors list of 1997, Dr. Meadow was knighted for "services to paediatrics and to the Royal College of Paediatrics and Child Health." His star was on the rise, and he became one of the most called-upon medical experts in trials of mothers in all of Great Britain.

At these trials, he would use the witness stand as a podium to promote his views, displaying a distinct talent for catchy phrases that the press loved to quote, such as "there is no evidence that cot deaths run in families, but there is plenty of evidence that child abuse does," or "one cot death is a tragedy, two is suspicious, three is murder." His views gained incredible notoriety, and on the strength of his highly respected word some 250 mothers were sent to prison.

UP ON the witness stand at Sally Clark's trial, Meadow was eager to share his knowledge and experience, and the conclusions he drew from them, with the judge and jury. Statistical studies showed that "the chance of a cot death in a family of the social status of the Clark family is about 1 in 8,543," he explained in his warm voice. "That means that the chance of two such deaths occurring in the same family is equal to the square of that number: one chance in about 73 million."

Sally's counsel begged to differ. Records from the CONI program, which followed babies born after a SIDS death in a family, showed that of five thousand babies monitored, eight had died. Surely that proved that the probability of such an event was much higher than 1 in 73 million, since the latter figure predicted that a double crib death would occur in England about once in a century. The CONI statistics showed that in reality this sad event actually occurs in England every couple of years. Indeed, the Clarks received many letters of support from families who had lost two, sometimes even three children, to SIDS.

Yes, but the CONI program data, explained Meadow, had not been collected with the kind of scientific precision and standards of a properly conducted study. The figures that he was using, by contrast, came from the CESDI report, which was more comprehensive than the CONI information, calculating the number of crib deaths in various sectors of the population.

The CESDI report, whose full title was "Confidential Enquiry into Stillbirths and Deaths in Infancy," was a controlled study commissioned by the British Department of Health. In it the author, Peter Fleming, a professor at Bristol University, identified three major risk factors for crib death: a smoker in the family, an unemployed parent, and a mother under the age of twenty-six. The study provided probabilities for the occurrence of SIDS in the presence of one or more of these factors and in the absence of all three. In families where all three risk factors were present, the probability rose to 1 in 214; in families where all three factors were absent, it fell to 1 in 8,543, the figure cited by Meadow. The overall figure is a probability of 1 in 1,300. Thus, the figure of 1 in 73 million only concerned families of the Clarks' habits and income; in the overall population Meadow would have expected the chance of a double crib death to be 1 in 1,300 squared, or about 1 in 1.5 million, fifty times more than the probability he was using for Sally. In other words, Meadow agreed that there could be a legitimate double crib death about once every couple of years in England, corresponding to observed fact. But these deaths would typically not occur among people like the Clarks; such a family, according to his reasoning, would suffer a crib death only about once in a century. It was just too unlikely an event: the Clarks did not have any of the three main risk factors for SIDS, so why should both of their babies die by pure chance?

As a matter of fact, the CESDI study makes it very clear that other family factors exist that may affect the probability of SIDS, but are not yet understood. Meadow's essential error was to ignore this observation and to treat SIDS as a phenomenon that befalls babies as a consequence of completely random chance, like the lottery.

"So when Harry was born, the chance of his being a cot death was the same as Christopher's? One in 8,543—like tossing a coin? It's the same odds each time? Heads or tails?" questioned Sally's lawyer.

"It's the chance of backing the long outsider at the Grand National," replied Meadow, calmly displaying a horrific lack of taste. "Let's say it's an 80 to 1 chance you back the winner last year and next year there's another horse at 80 to 1 and you back it and it wins. To get odds of 73 million to 1 you have to back that 80 to 1 chance, four years running. It's the same with these deaths."

This response clearly indicates that Meadow viewed crib deaths as a random occurrence—in the face of the very CESDI study he was quoting, which warns that there may be unknown factors, even genetic ones, that increase the risk in certain families. The choice of the figure of 1 in 8,543 for the probability of a crib death occurring once in a family like Sally's is correct, since the number is obtained by observation of millions of families. But Meadow's calculation of the probability of two crib deaths by squaring that number relies on the totally unjustified assumption that crib death is a purely random event. If, in fact, there is a genetic trait that can cause crib death, then two crib deaths in a family may both be traceable to this trait and thus not be independent at all. Meadow's calculation is an example of Math Error Number 1: multiplying non-independent probabilities.

So why did Roy Meadow treat crib death as a random occurrence? When you think about it, it really makes no sense to note that there are factors that increase the risk of crib death while at the same time considering each occurrence of it as being absolutely random. It has to be one or the other; it can't be both. If it is random, it will strike independently of any risk factors. But the CESDI study clearly shows that that is not the case. If there are known risk factors, then there can be unknown ones as well; in fact there almost certainly are some above and beyond the three identified in the CESDI report, probably several.

Moreover, crib death is not a single phenomenon, but an umbrella term used to describe infant deaths that are not medically understood. These deaths, in fact, do have causes—it's just that doctors have been unable to ascertain them. It sometimes happens that explanations arise later on, because of genetic features that continue to arise within the family or upon a more serious examination of the autopsy records. Once the cause of death is known and identified, the baby has no longer died of SIDS, and the statistics concerning SIDS are modified by the removal of the case from the databases. SIDS is not an absolute event that either has or has not occurred, nor is it a purely random event, so multiple occurrences cannot be assumed to be independent events. Unfortunately for the Clarks, however, Meadow's figure was accepted without question by judge and jury.

And not only was it accepted, but also it was misconstrued. The second problem with a figure such as 1 in 73 million is that even if it were correct, it tends not to be understood correctly. The public, and no doubt many members of the jury, took it to be the probability that Sally Clark might be innocent—that there was, in fact, a chance of just 1 in 73 million that she might be innocent. In other words, the reasoning was as follows: "Such an event happened to Sally Clark, there's a chance of just 1 in 73 million for that event to happen naturally; therefore it is practically certain that it did not happen naturally; Sally Clark must have made it happen."

This logic, which is almost irresistible, is also wrong (another example appears in "the incredible coincidence" discussion of Math Error Number 7 in this book). The fallacy becomes immediately obvious in the analogous statement: "One million lottery tickets were sold and Mr. X won; there was only 1 chance in 1 million for that to happen naturally, so the probability is too low to believe that it happened naturally; therefore Mr. X must have cheated." Of course, in lottery situations we know this is not true; someone always wins the lottery, and no one suspects the lucky person of having cheated.

The point is that double crib deaths, while extremely rare, do happen, and some unfortunate family somewhere will fall victim to it, just as somewhere there will be a fortunate Mr. X who wins the lottery. Once the event has already occurred, you cannot retroactively calculate the probability that it could have happened and then suspect that the likelihood is too small for it to have really happened. When it's the lottery, no one ever has a doubt.

In addition to the possibility that Sally's babies died by pure chance and the possibility that she murdered them, there was a third possibility, by far the most likely of the three: *that they died of an actual medical cause that the doctors had been unable to determine.* But the jury members were never told this. They were left to choose between "1 chance in 73 million that it happened by chance" and "otherwise she killed them." How could they hesitate?

Sally Clark was convicted of murder by a 10–2 majority verdict on November 9, 1999, and given the mandatory sentence of life imprisonment. The press had a field day reviling her as a child murderer. On her arrival at Styal Prison she could hear the other prisoners, who had followed the news on television, screaming, "Here comes the murderer!" and, "Die, woman, die!" as they strained and clambered to get a better look at her.

Thanks to "1 in 73 million," Sally Clark had suddenly become the most hated citizen of Great Britain. Nearly a year later, on October 2, 2000, the Court of Appeal upheld her conviction. They denied the influence of the statistic on the jury, writing, "The point on statistics was of minimal significance—a sideshow—and there is no possibility of the jury having been misled."

A request to the House of Lords for leave to appeal again—Sally's final chance for justice in Great Britain—was rejected. She faced life in prison, without even the hope of early release, which would have been possible only if she accepted to admit guilt and express remorse. But Sally was innocent. Not for freedom—not even for her life—would she say that she had killed her sons.

THE ONLY ray of hope in the months that followed was that the Family Court granted Steve Clark full custody of their remaining son, making it possible for Sally to see the child each week and even spend a full day with him once a month. Steve sold Hope Cottage, moved near where Sally was imprisoned, and devoted himself to continuing the struggle for her freedom while learning to be a single stay-at-home dad. Steve's professional and family life had been shattered, and nothing remained to him but his little boy.

> He and I become a team—he is my little mate, and we develop a strong bond of love; I get closer to him than most dads, but why did it have to be like this? We manage to muddle our way through together. Sometimes I sit outside his nursery all night, just in case he needs me . . . Then comes the morning I take him for his first day [at nursery school]. I cannot bear it. I cannot handle the thought of leaving him with strangers. But we walk there together, he holding my hand, sometimes quite tightly. Suddenly, much too soon, we are there. I don't want him to see me crying. But I can't help it. I kiss him goodbye and hand him over to a lovely lady, tears coursing down my face. I cry all the way back to the house; it feels strangely empty. What have I done? I sit there, desolate, terrified that something may happen to him.

Fortunately, Steve refused to give up. With the help of the lawyer who had worked on Sally's case from the beginning, and of many other people who generously gave their time for free, he continued to chase up every avenue that might lead to anything—an appeal to the European Court of Human Rights, a submission to the Criminal Cases Review Commission, a public relations specialist to help the truth filter out to the public, and as a last resort, further analyses of the medical examinations performed on Christopher and Harry, the results of which were being kept under lock and key by the hospital where they died.

The medical results had not initially been at the top of Steve's priorities. He was convinced that the experts had seen everything there was to see. Struggling with bills and a time-consuming new job, Steve had other things on his mind. However, as the case's exposure grew, people appeared out of nowhere to help the cause. One of these volunteers was a lawyer who wanted to obtain the medical records from the hospital, convinced that Steve's team needed them. Among other things, they wanted to get the original apnea alarm that had often rung when Harry was wearing it. They thought it possible that the alarm was not defective at all, and that in fact the child had undergone repeated episodes of abnormal apnea that had not been recognized by the health professionals checking on him.

Instead, after months of legal efforts, when the records were finally, reluctantly made available, the team found something else—something completely different, and shocking; something that had been overlooked by every single doctor who had had direct access to the records, meaning every single medical expert for the prosecution. Not by the defense, though. As a matter of fact, these documents had never been disclosed to the defense.

No fewer than eight different colonies of the lethal bacterium *Staphylococcus aureus* had been found in Harry's body, some appearing with polymorphs, which are the cells that our bodies develop to fight off disease. They showed that the baby had been suffering from a serious bacterial infection when he died, one that even could have led to meningitis. Confronted with these records, a dozen new and independent medical experts wrote reports stating that Harry most certainly could have died, and very probably *did* die naturally, from a serious infection. His death never should have been considered an unexplained crib death.

At around the same time, Meadow's mathematical assertions were put under scrutiny when on October 23, 2001, the Royal Statistical Society sent a public complaint to the Lord Chancellor, in which they exposed his errors and harshly expressed their gravity.

This approach is, in general, statistically invalid. It would only be valid if SIDS cases arose independently within families, an assumption that would need to be justified empirically. Not only was no such empirical justification provided in the case, but there are very strong a priori reasons for supposing that the assumption will be false. There may well be unknown genetic or environmental factors that predispose families to SIDS, so that a second case within the family becomes much more likely.

All of these facts were added to Sally's file when it came up in front of the Criminal Cases Review Commission, and her conviction was quashed on January 29, 2003. Sally was finally free. But she had spent more than three years in prison, and in spite of the joy of being reunited with her husband and child, she found it terribly difficult to recover the habits of a normal life. Having been accused of murdering her children because she was obsessed with her career, she could not contemplate going back to work. Having been told that she killed Harry because he was messy and disruptive, she cringed when friends admired her tidiness. Everything she had been, everything she was proud of in her life, had been held up as a model of horror to the entire country. And on top of this, she had been deprived of the ability to make a single decision for herself throughout her years in prison.

Sally suffered from a recognized psychological phenomenon known as "enduring personality change after catastrophic experience." In desperation, she sought consolation in alcohol, as she had for a short period after Christopher's death. She died of acute alcohol intoxication in her home on March 16, 2007, just four years after her release. She was forty-two years old.

THE TESTIMONY of Roy Meadow, recognized expert on Munchausen Syndrome by Proxy, child abuse, and the evils of motherhood, sent dozens of mothers to prison. After Sally's successful appeal, other cases were quickly forwarded on to the Criminal Cases Review Commission, and other mothers were exculpated and freed. One of them was Angela Cannings.

Angela had lost not two but three babies to inexplicable sudden death, the third literally days after Sally's first conviction. Although there was a history of infant death in part of Angela's family, which may have indicated an unknown genetic factor, she was accused of murder after the third occurrence and brought to trial. Sir Roy Meadow was the key expert witness for the prosecution. No one has described Meadow's particular style on the stand as powerfully as Angela did in her book *Against All Odds,* which tells the story of her tragedy. Quoting some of his testimony, she writes:

> I remember one exchange late in the day which made me shudder. Mr. Mansfield [Angela's lawyer] yet again insisted that looking at the whole picture—me, our family, the lack of injuries on Jason and Matthew, the features consistent with cot death—it was a real possibility that my children could have died from natural, but as yet unknown, causes.
>
> "I think the problem with that statement [is that] Mr. Mansfield is saying because the family is normal, child abuse doesn't happen," Professor Meadow replied. "It is absolutely right to say that child abuse and smothering are more common in certain families, but nevertheless, most abuse, most smothering happens in families who on ordinary meeting seem normal and caring and that is so, and most of the mothers who smother children, when you meet them, are normal. The second point is to start talking about the features of SIDS. SIDS means that you don't know why the baby has died. It means that an unnatural cause such as smothering wasn't found, and nor was a natural cause, so that in any group of SIDS babies there are some who have been smothered."
>
> I was trapped. If I appeared normal, I could be a child abuser; if my babies were thought to have died of cot death, I could have smothered them. There might not be any actual proof against me but Professor Meadow had created a world of smoke and mirrors from which I could not escape.

Like Sally, Angela was sentenced to life in prison. She immediately appealed. While she was waiting, she heard about the trial of Trupti Patel, another mother who had lost three babies to SIDS. Trupti's trial took place six months after Sally's release. Meadow testified against her, too, and listed no fewer than four different reasons proving that she must be guilty of murder. Fortunately for Trupti, however, the stinging judicial criticism of Meadow's errors that had been published at the quashing of Sally's guilty verdict had rippled out into public consciousness by then, and when other possible causes for the death of Trupti's babies were discussed—a genetic defect in particular—the jury listened carefully and acquitted her. Following this, the Solicitor General of England and Wales effectively barred Sir Roy Meadow from testifying for the prosecution in any further trials. Angela Cannings' conviction was overturned just months later, and following her release, cases of imprisoned mothers elsewhere were reviewed. But as was the case for Sally, it was too late for Angela to put the broken pieces of her family life back together. As she described movingly in her book, finding her husband sunk in depression and her one remaining daughter in a state of psychological disturbance, Angela struggled for many months before finally leaving her home to try to begin a new life.

In July 2005 the British General Medical Council (GMC) found that Sir Roy Meadow had been guilty of serious professional misconduct in his misuse of statistics at the trial of Sally Clark, and his name was struck from the medical register. The GMC's decision was later overturned and Dr. Meadow was reinstated, but by that time he had retired.

Meadow always denied any wrongdoing, admitting only that what he had done was perhaps "insensitive." But the GMC termed his actions "fundamentally unacceptable." The panel stated that while Sir Roy Meadow was recognized as an eminent pediatrician, "he should not have strayed into areas that were not within his remit of expertise." His calculation could be valid only if it were known that two crib deaths within a family must necessarily be independent of each other, but in fact there is no known medical justification for that assumption and many reasons to believe it false.

Meadow stood by his reasoning. But he regretted having used the example of betting at the Grand National to illustrate the probability. It was the only thing he regretted, apparently. None of the innocent mothers who spent years in prison because of him, none of the families whose lives were shattered, ever heard a word of apology.

MATH ERROR NUMBER 2 »
UNJUSTIFIED ESTIMATES

IT IS DIFFICULT TO OVERESTIMATE the extent to which we are bombarded with figures on a daily basis. Intended to inform us, to enlighten us, to help us, these figures also, far more frequently than one might like, mislead us. A shocking proportion of the numerical estimates we receive are simply wrong, whether by intention, by accident, or because of ignorance or typographical error. Worse, the effect of this kind of error is frequently minimized, as though the only important thing is having a number at all, one that can lend a scientific aura to whatever statement is being made.

A British report on the failings of the Labour government from February 2010 cited a figure of 54 percent for the proportion of girls in the ten most disadvantaged areas of England who became pregnant before the age of eighteen. When challenged by an alert reader who realized that this figure seemed unreasonably large, the Tories admitted that the correct estimate was actually 5.4 percent. The error would have been forgivable if the Tories had not felt it necessary to make the following public statement: "A decimal point was left out in a calculation. It makes no difference at all to the conclusions of a wide-ranging report which shows that Labour have consistently let down the poorest in Britain."

This *we're giving you a number but who cares whether it's right or wrong* attitude ends up weakening our capacity for making our own assessments, because, after all, if it doesn't matter, then why bother giving figures at all?

But it does matter. In the next case, not only were the statistical figures given in court multiplied together incorrectly as in Math Error Number 1, but also the figures themselves were inaccurate estimates thrown out to the jury by an enterprising prosecutor. Once caught, these errors eventually led to the overturning of a remarkable conviction—but not before the accused had already completed their terms in prison.

The Case of Janet Collins: Hairstyle Probability

Juanita Brooks crashed painfully to the ground, her cane underneath her. The groceries from her wicker shopping basket scattered over the pavement. Stunned and in pain, it took the elderly woman a moment to realize that she had been violently pushed from behind, and another to lift her head and scan the area for her attacker. What she saw was a young blonde woman tearing down the alley and rounding the corner at the far end. Dangling from her hand was Juanita's purse.

John Bass, who lived on a street off the end of the alley, was outside watering his front lawn when he heard Juanita scream. As he looked up from his hose, he saw a young woman, blonde ponytail flying behind her, run out of the alley and jump into a bright yellow car that was waiting at the curb. The car revved up and took off, swinging widely around a parked car and passing within six feet of Bass. To his surprise, he noticed that the driver was a black man.

It was 1964. Interracial couples were very rare, and they were not treated with indifference. In public, in the street, they were noticed, singled out, and frowned upon.

The investigator assigned to the case was Officer Kinsey of the Los Angeles Police. Kinsey collected as many descriptive details from the two witnesses of the crime as he could. From Juanita he learned that the woman she had seen appeared to be generously built—Juanita estimated her weight at 145 pounds—with hair "between dark and light blonde." She also described the woman's clothing as "dark." Bass agreed about the dark clothing and guessed that the woman he had seen was about five feet tall, but he described her build as "ordinary" and mentioned that her dark blonde hair was tied back in a ponytail, which Juanita could not remember. He also stated that the driver of the car wore a mustache and beard.

The police had no clues to the identity of the bag snatchers and no clear trail to follow. Juanita Brooks' son, however, was enraged by what had happened to his mother: not only had she been attacked and robbed, but also she had sustained a dislocated shoulder from her fall. He was determined to find the attackers himself. Having come up with a simple plan, he visited every gas station in the neighborhood with a description of the pair until he hit on one whose personnel confirmed that indeed an interracial couple came there regularly to fill up the tank of their yellow Lincoln. Brooks took this information straight to the police, which explains how, four days after the robbery, Kinsey was ringing at the doorbell of the modest house inhabited by Malcolm and Janet Collins.

When the police officer chose to follow up the information that Juanita's son brought him—when he agreed that the Collinses were suspects simply because they matched the description of the thieves—he could not have realized that he was engaging in an unorthodox identification procedure that would later result in a serious legal puzzle. He assumed he would probably find evidence of their crime easily enough; maybe he would even obtain a confession.

As Kinsey drew up in front of the Collins house, the first thing he saw was a yellow Lincoln parked in the street, and when Janet answered the door, he noted with satisfaction that she was wearing a ponytail. It was blonde, though he would have described it as light rather than dark, and Malcolm was not wearing a beard, but there were enough features in common with the thieves as described by Juanita and Bass for Kinsey to feel justified in asking the Collinses to accompany him to the police station. There he had them photographed, and interrogated them about their activities at the time of the robbery. Janet explained that Malcolm was unemployed, but that she had been at work at her job as a housemaid in San Pedro on the morning of the robbery starting at 8:50 a.m., and that her husband had come with the car to pick her up at 1:00 p.m. According to both of them, they had then driven to the home of a friend in Los Angeles and had spent the whole afternoon there. Officer Kinsey released them and had them driven home in a police car pending further investigation.

His next step was to show the photographs of Malcolm and Janet to the victim and to his only witness. The result, however, was disappointing. Juanita could not identify Janet at all, and Bass was able to say only that the ponytail of the woman he had seen running away "looked the same" as the one in the picture. However, Kinsey was not deterred. He was pretty certain that Malcolm Collins was a shady character. And he had a plan to prove it.

A DAY or two later, Officer Kinsey drove around the area where Malcolm and Janet lived until he saw them arrive home in their yellow Lincoln. He followed them, parked his car in a position from which he could survey the rear of the house, and put in a call for backup at the Collins home. Following his instructions, the additional officers rode in a marked police car and pulled up ostentatiously in front of the house. The car immediately disgorged numerous uniformed officers, who stormed up the front path and loudly rang the doorbell.

This tactic produced exactly the effect that the officer had hoped to achieve. From his position at the rear, Kinsey saw Malcolm go racing out of the back door, scurry into a neighbor's garden, and enter the house. When Janet opened the front door, the other officers went straight into the Collins home, arrested her, and led her outside, where they pushed her into the police car. Then they went next door, entered the house, and began a room-to-room search for Malcolm, whom they discovered tightly squeezed inside a closet. The couple was taken back to the police station, and this time they were kept in custody and interrogated for more than forty-eight hours. Yet once more, in spite of their shocking treatment of the suspects, the police failed to obtain a confession or even a shred of solid evidence, and for a second time Malcolm and Janet were released with no charges.

Frustrated by the failure of his second premature attempt to extract a confession, Kinsey set to the tedious work of gathering evidence, questioning Janet's employer, the friend whom the Collinses had visited on the day of the robbery, and several of their acquaintances and neighbors. He collected as much information as he could—concerning the Collinses' alibi, their financial situation, their hair color and style, and the past and present state of Malcolm's beard—and two weeks later he arrested Malcolm and Janet for the third time.

Although the facts that Kinsey collected through his investigation were actually rather imprecise and confusing, he decided that this time they were sufficient to justify the arrest, on the basis of two rather strong points against the couple. First, although Janet had claimed that her husband had picked her up from work at 1:00, her employer stated that the time had actually been about 11:30. True, 11:30 was the time at which Juanita said she had been robbed, which would have made it impossible for Janet to be the perpetrator. But neither Juanita nor Bass had been able to give the time of the robbery with any real precision, and the distance between Janet's employer's home and the robbery could have been covered in just a few minutes by car.

Second, on the day following the robbery, Malcolm had paid off two traffic fines totaling thirty-five dollars. The police found the receipts in his pocket. Asked separately about the funds used to pay the fines, Malcolm explained that he had used money he had won at a gambling hall, and Janet said he had paid them out of some of her savings. The trouble was that the stolen purse had contained a sum of money that Juanita estimated as between thirty-five and forty dollars.

To get an idea of what that sum represented in 1964, it is useful to have a few points of comparison. The average monthly rent in the United States that year was $115; a loaf of bread cost around $0.20. Janet Collins earned $12 a week for her part-time work as a housemaid. On the one hand, her $50 per month might seem hardly enough to survive on, but on the other, certain things cost unimaginably less than they do today. Janet and Malcolm had gotten married on June 2, two weeks before the date of the robbery, and had taken a honeymoon to Tijuana, Mexico. The trip cost Janet only "a part of that $12."

Still, at Janet's part-time salary, those traffic fines must have represented a heavy financial burden for the Collinses. At the same time, the money obtained from the bag snatching would have seemed like a near-miraculous piece of good fortune. Even if they had been so desperate that they had planned to snatch someone's purse, they could hardly have imagined finding a sum as large as thirty-five dollars—exactly what they needed to pay their fines. Remember, that amount of money would have been nearly Janet's monthly salary; not many people walked around carrying so much. The matching of the sums seemed almost too good to be true.

WHILE IN custody and awaiting the preliminary hearing, Janet grew afraid, and requested a private conversation with Officer Kinsey. During this conversation, Janet told Kinsey of her fear that because her husband had a previous criminal record, a new conviction would net him a longer prison term. She repeatedly expressed her anxiety on this subject, and finally she told Officer Kinsey that if either of them were to be convicted, then it should be her and her alone; she wanted to take all the blame. In excerpts from the conversation, which was recorded in its entirety and played back during the trial, she asks about this possibility again and again.

"If I told you that he didn't know anything about it and I did it, would you cut him loose?"

"I just want him out, that's all, because I ain't never been in no trouble. I won't have to do too much time, but he will."

"What's the most time I can do?"

"Would it be easier if I went ahead and said, if I was going to say anything, say it now instead of waiting till court time?"

At a certain point, Officer Kinsey summoned Malcolm to join in the conversation. Perhaps he hoped to extract some contradictions between their stories—for example, about exactly what money was used to pay the fines—and somehow parlay those contradictions into a confession. But that didn't happen. Instead, like Janet, Malcolm seemed interested only in discussing strategy to obtain the least possible total punishment. "I'm leaving it up to her," he can be heard saying at one point in the recording. And at another: "This is a little delicate on my behalf."

The conversation ended when the couple finally decided that they needed more time to think things over. To Officer Kinsey the tone of the discussion and the concern of the couple with their possible prison sentences were signs of guilty behavior, and he described them as such during the subsequent trial, saying, "They seemed to be conscious of their guilt, and looking to find the best solution to get out of the situation."

But just as Malcolm Collins' fleeing the police to hide in a neighbor's closet may not indicate actual guilt so much as a general fear of trouble with the police due to bitter past experience, so this conversation can be understood in quite a different manner. Married to a black man at a time of rampant racism, nineteen-year-old Janet must have repeatedly experienced the censure of society. Ignorant, working-class, and poor, she would not have been used to standing up for her rights. She probably felt that, guilty or innocent, there was essentially no chance at all that a trial would result in acquittal. Her marriage to an unemployed black man with a previous record fit into a perfect image of a couple of petty criminals, and jail time must have looked like a certain bet. Under those circumstances, Janet's desire to shoulder the blame in order to save her husband from more serious trouble may seem indicative not so much of guilt as of love.

AS WAS revealed over the course of the couple's joint seven-day trial, the prosecution's entire case was built around the problem of identification. The defense stressed the impossibility of someone leaving work at 11:30 and committing a crime some streets away, also at 11:30. The prosecution responded by noting that none of the witnesses, neither Juanita, Bass, nor Janet's employer, were certain of their times to within a few minutes. And as noted above, that was all it would have taken for the couple to drive to the scene of the robbery.

There was also the issue of Malcolm and Janet's alibi. Unfortunately for them, although the friend whom they claimed to have visited in Los Angeles on the day of the robbery remembered their coming over to her place, she was not able to recall the precise time or even the precise date of their visit.

Still, both of these pieces of evidence were weak; they didn't prove anything. What the prosecution wanted was to pinpoint Malcolm and Janet Collins as the robbers by identifying them through the physical evidence given in Bass' description: the yellow car, which they undoubtedly did possess; Janet's dark blonde hair, ponytail, and dark clothing; and Malcolm's mustache and beard.

If those details had been more precise and corresponded more closely to the two defendants, they might well have been in serious trouble. But in fact, they were vague and did not quite fit. For one thing, Malcolm was not now wearing a beard. Bass nonetheless identified Malcolm at the trial as the man he had seen in the car. But the effect of his declaration was destroyed by the defense, who provided evidence proving that shortly before the trial Bass had failed to pick Malcolm out of a police lineup.

Asked whether he had worn a beard on June 18, the day of the robbery, Malcolm explained that although he did occasionally wear a beard, he had not been wearing one lately, having shaved it off a couple of weeks earlier for his June 2 wedding to Janet. The defense called a number of witnesses who were acquainted with Malcolm and confirmed his claim. However, the court clerk who took payment for Malcolm's traffic fines on June 19 testified that he recalled him as having a beard on that day. In the end, this point simply could not be determined one way or the other.

Then there was the problem of Janet's clothes and hair. Evidence was presented that on June 18 she had been wearing light-colored clothing, not dark. Furthermore, neither Juanita nor Bass was able to identify Janet in the courtroom. On top of that, while they both agreed that Janet's hair seemed somewhat lighter in court than it had been on the day of the robbery, Janet's employer testified that she thought that Janet's hair had become darker since that day. The possibility that Janet had dyed, bleached, or otherwise altered her hair since June 18 was discussed in detail, but could not be resolved. Janet herself denied having done anything special to her hair.

Malcolm and Janet were both called to testify on their own behalf, and both denied any involvement in the crime. Malcolm told the jury about how he had picked up his wife at work and driven her to their friend's home in Los Angeles, where they spent the afternoon. Janet confirmed this. Of her conversation alone with Officer Kinsey, she added on the stand that "inducements had been held out to her on condition that she confess her participation," and she formally denied ever having made any confession or intending to do so.

There were no further witnesses, and therefore no further testimony was expected. It was obvious that the case against Malcolm and Janet was weak. But at the last minute, as a sudden move to bolster up the failing process of identification, the prosecutor presented a dramatic new approach.

RAY SINETAR, a thirty-year-old prosecutor with just two years of experience behind him, had been wondering how to explain to the jury what he saw intuitively: the Collinses had to be the thieves for the simple reason that the number of couples fitting their description was so small that, it seemed to him, they were virtually certain to be the only such couple in the neighborhood. It was frustrating to perceive this so clearly and yet not have the evidence to make it hold water in court. But Sinetar had a little knowledge of mathematical methods; his brother-in-law was Ed Thorp, a mathematician and blackjack genius from New Mexico who would testify in the role of expert witness two months later at the murder trial of Joe Sneed (see chapter 3). It struck Sinetar that maybe he could mathematically *prove* that the Collinses had to be the right couple, by calculating the probability that any given couple in the Los Angeles area could share their distinguishing features, vague as they were: mixed race, yellow car, mustache, beard, blonde ponytail.

Early in the morning of the second day of testimony, Sinetar dashed off a phone call to the local university, California State University at Long Beach, and left a message that he urgently needed a mathematician to come into court and testify. The man who answered the message was twenty-six-year-old probability theorist Daniel Martinez, who had come in to work that day to teach his class and thought it might be interesting to turn his knowledge into evidence in a court case. He later recalled, however, feeling a bit unsettled when the case began to unfold before him, wondering just what he had got himself into.

In court, Sinetar turned to the jury and explained to them that he was going to give a mathematical proof. He would show that if a person searched for couples matching the physical characteristics described by the witnesses to the crime—couple in car, black man, beard, mustache, white woman, blonde, ponytail, yellow car—the possibility of a precise match was so unlikely that if any such couple were actually to be found, the odds would be overwhelming that it must be the same couple as that seen by the witnesses.

Sinetar put Martinez on the stand and had him testify to the validity of the product rule, which, as we saw in the previous chapter, states that *if two events are independent, then the probability that they both happen is obtained by multiplying the probabilities for each one happening on its own.*

Next, Sinetar gave estimates for the separate probabilities associated with finding a person with each of the couple's distinguishing characteristics, as follows:

- Black man with a beard: 1 out of 10
- Man with mustache: 1 out of 4
- White woman with blonde hair: 1 out of 3
- Woman with a ponytail: 1 out of 10
- Interracial couple in car: 1 out of 1,000
- Yellow car: 1 out of 10

Sources differ as to whether Sinetar gave Martinez the probability figures without mentioning the qualities they corresponded to, or whether he told Martinez that judging the validity of those probabilities was not his concern. In a telephone interview some forty years later, Sinetar recalled that he only gave the figures, but a quoted question and answer from the original testimony recorded in the state supreme court appeal judgment indicates that he said more:

> Now, let me see if you can be of some help to us with independent factors, and you have some paper you may use. Your specialty does not equip you, I suppose, to give us some probability of such things as a yellow car as contrasted with any other kind of car, does it? . . . I appreciate the fact that you can't assign a probability for a car being yellow as contrasted to some other car, can you?

Martinez's recorded answer: "No, I couldn't."

In any case, what emerges clearly is that Martinez was asked to multiply the numbers together and was provided with paper and pencil for the purpose. He did so, obtaining the result $\frac{1}{10} \times \frac{1}{4} \times \frac{1}{3} \times \frac{1}{10} \times \frac{1}{1,000} \times \frac{1}{10} = \frac{1}{12,000,000}$, a probability of 1 in 12 million.

Having made this calculation, the mathematician was asked to stand down, and the prosecutor gave his interpretation of this result to the jury in an impassioned speech. He explained that this chance of 1 in 12 million represented the likelihood of a given pair of persons in Los Angeles fulfilling all of the criteria above—being seen together in a car, having a blonde ponytail, etc.—and that therefore, having found such a couple, one could be sure far beyond any reasonable doubt that this must be the couple in question. In fact, the prosecutor told the bemused jurors, this new type of mathematical proof was on the verge of replacing the traditional idea of proof beyond a reasonable doubt, a notion he described as "the most hackneyed, stereotyped, trite, misunderstood concept in criminal law."

Recognizing that some people might be disturbed by the replacement of an actual search for solid proof of guilt by a purely theoretical and numerical operation, and that this might in fact even be a source of judicial error, he admitted that "on some rare occasion . . . an innocent person may be convicted." But that can happen anyway, he said, and if it came to a choice between using the "new math" to convict the occasional innocent or using the old system and letting the guilty go free, then surely the new math was preferable, for otherwise, "Life would be intolerable . . . because there would be immunity for the Collinses, for people who choose not to be employed to go and push old ladies down and take their money and be immune, because how could we ever be sure they are the ones who did it?" Sinetar concluded his speech by noting that the estimates he had given were actually conservative, that the real values were probably even smaller, so that "in reality, the chances of anyone else besides these defendants being there . . . is something like one in a billion."

It took the jury eight hours and five ballots to deliver a verdict of guilt. Surprisingly, a case of little importance in the annals of crime turned out to have tremendous significance in the annals of law. In a way, what the jury was doing that day was not determining whether the Collins couple was guilty or not. Rather, they were making a judgment about whether mathematical calculation could replace hard evidence. There was no serious evidence against the Collinses: nothing about their appearance could be identified clearly, their alibi was not firm but could not be proved false, and as for their poverty, miserable social situation, and fearful behavior around police, not only were these factors insufficient proofs of guilt, but the third one may well have been caused by the other two.

The jury found the Collinses guilty of second-degree robbery, and they were sentenced to prison terms. The public reactions to Sinetar's legal exploit followed quickly. "Justice by Computer" and "Law of Probability Helps Convict Couple" were samples of the headlines that appeared soon after the trial. Attention started to build up around the event, and within a month a feature on the case ran in *Time* magazine. In its issue of January 8, 1965, under the heading "Trials: The Law of Probability," *Time* told the nation:

> A jury has convicted the [Collins] couple of second-degree robbery because Prosecutor Ray Sinetar, 30, cannily invoked a totally new test of circumstantial evidence—the laws of statistical probability.
>
> Convicted by math, Malcolm Collins received a sentence of one year to life. Janet Collins got "not less than one year."

JANET COLLINS did not appeal her sentence, preferring to see it through, stay under the radar, and keep out of the way of the law in future. But Malcolm, perhaps more rebellious by nature, and possibly encouraged by a lawyer who sensed an opportunity for legal innovation, did appeal. And when he lost, he appealed again, and the case went up to the supreme court of California. There, Sinetar's technique met its match—in the person of twenty-five-year-old law clerk Laurence Tribe, who was assisting one of the court's judges.

It so happened that Tribe had majored and excelled in mathematics at Harvard before turning to law school. Thus, in his memo for the judge, he was able to systematically call out all the errors that Sinetar had managed to make in the course of his misleadingly simple argument. Readers of the state supreme court's judgment will find Tribe's memo, unsigned, at the end, and any mathematician will recognize at once that it was written by an unusually knowledgeable hand. Tribe's arguments are flawless and convincing.

The first two errors he raised are none other than Math Error Numbers 1 and 2. The prosecutor gave arbitrary values out of his own head for the probabilities of such features as cars being yellow—not to mention the absurd naming of a "probability" that a girl will be wearing a ponytail, given that hairstyle, unlike car color, can be altered instantaneously and at will. As for the assumption that the thieves were actually a married couple, it seemed to have no justification whatsoever; there was nothing to prove that the people in the yellow car at the scene of the robbery were married. In short, Sinetar's assumptions, and his numbers, were based on no statistical research (or almost none—Sinetar does recall having asked the law office secretaries for their guesses before making his chart!) and certainly no hard evidence. Such estimations can be valuable in everyday life; the ability to make an educated guess is one of the weapons people can use to fight against the abuse of numbers in the public domain. But vague estimations have no place in a court of law—no person's freedom should depend on them. As though to underline their inaccuracy, the prosecutor went so far as to tell the jury that he considered his estimates to be quite "conservative" and, as pointed out in the *Time* article, invited them to make up their own, suggesting that practically any numbers would do.

The supreme court judgment reads:

> The prosecution produced no evidence whatsoever showing, or from which it could be in any way inferred, that only one out of every ten cars which might have been at the scene of the robbery was partly yellow, that only one out of every four men who might have been there wore a mustache, that only one out of every ten girls who might have been there wore a ponytail, or that any of the other individual probability factors listed were even roughly accurate.

The prosecutor's next error was to ask his expert witness to apply the product rule to his probabilities, without verifying or allowing the witness to verify whether all the events in his list were independent. This was a serious mistake, since many of those events are actually not independent at all. The probability of a man having both a mustache and a beard is most certainly *not* equal to the probability of a man having a mustache multiplied by that of a man having a beard, given that beards not accompanied by mustaches are quite rare.

As a matter of fact, this flaw had not gone unnoticed. Rex deGeorge, the lawyer who defended Malcolm at his appeal, had already raised the same point, albeit unconvincingly. Who could take seriously his saying, "There is a dependency between Negro drivers and yellow cars; there are by far many more Negroes than Caucasians driving yellow cars," or, "There is a dependency between blondes and intermarriage; blondes and redheads tend to be more adventuresome, more daring, and more likely to choose to be with a Negro," or his claim that "there is a dependency between the way a woman would normally wear her hair and how she would fix it when she goes to carry out a robbery"?

DeGeorge's appeal failed, but his main point is justified, and was made more rigorously by Tribe in the supreme court brief.

> There was another glaring defect in the prosecution's technique, namely an inadequate proof of the statistical independence of the six factors. No proof was presented that the characteristics selected were mutually independent, even though the witness himself acknowledged that such condition was essential to the proper application of the "product rule" or "multiplication rule." . . . To the extent that the traits or characteristics were not mutually independent (e.g., Negroes with beards and men with mustaches obviously represent overlapping categories) the "product rule" would inevitably yield a wholly erroneous and exaggerated result even if all of the individual components had been determined with precision.

Other errors Tribe raised deal with more subtle difficulties. For example, he showed that even if one accepts that only one couple in 12 million shares the traits of the thieving couple, this figure cannot be confused with the probability that having found one such couple, it must be the right one. This is a different calculation and significantly more complex; we will face a very similar problem in the case of Diana Sylvester (see chapter 5). Indeed, Tribe calculated that the probability of another couple in the area existing with the same main identifying features as the Collinses (yellow car, blonde hair, etc.) was in excess of 40 percent.

Due to the misuse of mathematics and its excessive effect on the jury's decision making, the original judgment was reversed; Malcolm's conviction was quashed, and he was released to join his wife, who by that time had been out of jail for some three years already.

The bag snatching of which the Collinses were accused normally would have sunk without a ripple into the infinite sea of minor legal cases. But the novel use of mathematics in their trial shot it into the newspapers, and from there into the annals of legal history. Laurence Tribe went on to become a major name in American law, defending Al Gore against George Bush in front of the US Supreme Court after the 2000 presidential elections, teaching a series of brilliant students—including President Barack Obama, in whose administration Tribe later served—and, perhaps most importantly, writing a series of seminal papers in the early 1970s rejecting the use of mathematics at trial. It is practically impossible to overestimate the influence Tribe's papers have had in the criminal justice system, bringing the field of research into the proper and correct use of mathematics at trial to a virtual halt for decades. It would take a new generation, and all the mathematical difficulties associated with the modern science of DNA analysis, to revive the urgency of that movement and bring it back to life.

MATH ERROR NUMBER 3 »
TRYING TO GET SOMETHING FROM NOTHING

SUPPOSE YOU ARE PROOFREADING a one-thousand-page manuscript, and your work is supposed to be good enough to overlook no more than a maximum of twenty errors throughout the book. You have read fifty pages attentively, and so far you have not found a single typo. What could you conclude about the number of errors that you are likely to find throughout the manuscript? There are zero in your sample: how many might there be in the book?

Would you feel inclined to assume that the part you read is representative of the whole and estimate the number of errors as being fewer than one per fifty pages, or fewer than twenty throughout the whole book? Would you snap the pages shut, go to your boss, and assure him that the job is done?

Or are you the kind of person who would think of the 101 reasons why mistakes might still occur in the rest of the book? After all, it could be that a whole group of them appear together a bit further along in the text. The author might have had a bad day or an inattentive moment. He might have written an entire chapter on a topic that happened to be the only word he didn't know how to spell. Or maybe he typed the later text during a particularly bumpy train ride. If you're the latter kind of person, you wouldn't rest until you had thoroughly checked every single page; otherwise you'd never be certain that you had done a good job.

In fact, the best attitude depends on the exact work you are doing. If you are checking a product in which you feel certain that any errors will be distributed equally throughout—for example, an automatically produced product—you would probably be safe with the first method. But if you have no reason to believe that the errors will be spread out evenly, then you might easily miss something embarrassing. In the latter case, it's difficult to draw any conclusion about the total number of errors in the whole book based on the zero in your sample. Guessing that the whole manuscript is perfect would be risky—and making such a guess in a court of law is outright wrong.

The prosecution in the next case we describe combined the previous two math errors, giving unjustified statistical estimates and multiplying them together for non-independent estimates, and to top it off, the most important of these estimates was made based on a finding of zero in the sample examined.

The Case of Joe Sneed:
Absent from the Phone Book

The murder occurred on August 17, 1964, in the sweltering heat of a New Mexico high summer. Silver City, which had come into sudden existence as a tent city when a large silver mine was discovered there in 1870, had survived the plague of abandonment that turned so many of the nearby mining communities into ghost towns when the veins of ore ran dry. This was perhaps thanks to the splendid solitude of its situation in the midst of a southern desert paradise. Over the ensuing years, the town grew and developed into a ranching community, policed by sheriffs and, unofficially, by posses, and home to legendary criminal figures of the far West, such as Butch Cassidy and Billy the Kid.

The town was already blanketed by the cooling darkness of nighttime when Pauline Hicks, from the comfort of her home in the upper-middle-class section of Silver Heights, heard something that sounded like sharp shots ringing out into the shadows. Disturbed, she stepped out into her garden and walked over toward her neighbors' house to take a quick look

Silver City Mines (top), Butch Cassidy (left), and Billy the Kid (right)

around. All seemed quiet, however, so Mrs. Hicks returned indoors and went to bed. "I thought nothing more about it," she testified later.

But the next morning, August 18, a young man came running to her house, knocking and pounding on the door in a state of fear and shock. "Help!" he shouted. "My parents have been murdered! They've been shot!"

Mrs. Hicks recognized twenty-year-old Joe Sneed, her neighbors' son, although she had not seen him for some time. He had graduated from Silver City High School two years earlier, and had subsequently moved to California, where he had been living for about a year.

She and Joe called the police, who arrived almost at once and entered the house by the back door, which Joe had left open. They were shocked at what they found.

Ella Mae Sneed, Joe's forty-eight-year-old mother, was lying in her bed, dead, her head resting on her pillow. She had been shot three times: in the left ear, in the left side, and in the back. It was obvious that she had been killed in her sleep.

Her husband, fifty-year-old Joe Alvie Sneed, was lying in his pajamas in a small entryway between the bedroom and the bathroom, with a bullet wound to the side and two in the back. But the possibility that Joe had been killed because he was up and out of bed—roused by a burglar, perhaps, or in the heat of a quarrel—was quickly discounted. Bloody tracks as well as the position of his wounds proved that he, like his wife, had been shot in bed, probably while asleep. The poor man had succeeded in staggering to the door of the bedroom before collapsing on the floor, dead.

There was not the slightest sign of a robbery. Not even a forced entry. The bullets came from a .22 caliber pistol, which was not found at the scene.

The police brought young Joe down to the station to ask him for details about his discovery. The questioning was cordial; some of the officers were Joe's friends, and several had come to know him when he had worked in the streets of Silver City as a newspaper delivery boy a few years earlier. He told them that he had been returning to Silver City to visit his parents after his long absence. He had made the trip from California by car, and had spent the final night of his journey in a motel in Las Cruces, arriving at his parents' home early in the morning to catch them at breakfast. Discovering the crime scene on entering, he had rushed immediately to the neighbors' house for help without touching a single thing, he explained.

Asked if he was willing to take a lie detector test to check his story, he replied that he was. The test was administered at a specialized center in El Paso, and Joe's answers and reactions were closely monitored, but he remained perfectly coherent and quite unflappable, and the test indicated that he was telling the truth.

As a double precaution, police also subjected him to a paraffin test, which examines the skin of the hands for microscopic particles that become embedded there when the hands have recently fired a gun. But Joe's hands were absolutely clean. The police let him leave, and he went to stay with his grandparents in the nearby town of Central. On the following morning, August 19, a coroner's jury returned a verdict of death by gunshot wounds at the hands of a person or persons unknown. "Mystery mounts in double slaying of a prominent Silver City couple found dead in the bedroom of their home," reported the daily newspapers, "as lie detector and paraffin tests prove negative."

Joe Alvie Sneed and his wife were prominent citizens of Silver City. The couple was responsible for the circulation section of the *Silver City Daily Press*. They had two married daughters who no longer lived at home; young Joe was their third child and only son. Described in the local newspapers as "an average American youth," nothing emerged from his history in Silver City that might indicate any propensity to violent and shocking acts. Yet given that he was the discoverer of the bodies, the police were duty bound to treat him as a suspect. It seems, however, that in those first days they did not take this obligation as seriously as they might have; very probably they did not believe he was the murderer.

Unfortunately, this casual attitude led them to make a mistake—the first of a series—which nearly led to a grave judicial error.

ON THE day following the discovery of the bodies, Sergeant Richard Ingram of the Silver City police force called Joe's grandparents in Central and asked them to bring Joe in to the Silver City police station for further questioning. The young man drove up by himself, using his grandfather's car. Advised that he did not have to answer questions, he appeared surprised, saying that he really wanted to answer them, that he intended to cooperate with police and try to be useful, and that he absolutely wanted to know who had committed the crime. At some point he mentioned to the officers that he would like to get his own car back, since it had remained at his parents' house when the police had driven him to the police station the day before. Sergeant Ingram, who was heading the investigative team, told Sneed that he would go fetch the car. He claimed that Joe responded by willingly handing over the keys.

The police fetched the car, but that was not all they did. They also searched inside—without a warrant, although they could have obtained one within the hour—and discovered something rather suspicious there that gave them something specific to look for as they gathered their evidence in

Las Cruces. On the very next day, August 20, Joe Sneed was arrested for the murder of his parents. At the preliminary hearing, Sneed and his lawyer, J. Wayne Woodbury, contested the production of the documents found in the car on the grounds that the officers had obtained them by unlawful search and seizure. During the hearing, Sergeant Ingram was asked exactly what had given him the right to conduct such a search without obtaining a warrant. He claimed that he had asked permission from Joe himself, who had voluntarily given it.

INGRAM: Joe, the defendant, was worried about his car, and I told him that we would try to get it down to him as soon as we could . . . I went and asked Joe if we could have the keys to his car, that we wanted to search it and then we would bring it down to the City Hall.

Question: What did he say?
INGRAM: He handed me the keys.

Question: Was Mr. Sneed under arrest at that time?
INGRAM: No.

Question: For what purpose did you wish to question him at that time?
INGRAM: I had some questions I would like to ask him about his trips, et cetera.

Question: What prompted you to seek these questions, these answers from Mr. Sneed?
INGRAM: I was trying to find a clue.

Question: I see, in other words you were trying to find a clue against Mr. Sneed?
INGRAM: No, just a clue to shed a little light on the case . . . I didn't interrogate him, I just asked him some questions.

Question: Make some distinction for me, it is yours, what's the difference between asking questions and interrogating?
INGRAM: I wasn't accusing him of anything.

Question: Was he not then under arrest?
INGRAM: No, sir.

Question: Nor was he a suspect? No more than anybody else might be?
INGRAM: I guess more so than a lot of people.

Question: More so than a lot of people?
INGRAM: Yes.

Question: But so far as you were concerned, this was a friendly assistance Joe was lending to the police department.
INGRAM: Yes.

"Unlawful search and seizure" is the act of conducting a search with neither the permission of the owner of the searched property nor a properly issued warrant. In his testimony, Ingram indicated that Joe Sneed had given an oral assent to the proposition of searching the car, but it sounds somewhat forced, given that Sneed was not a suspect at the time and probably would have reacted with surprise if asked specifically whether his car could be searched. Joe himself denied having been told any such thing, and he certainly never waived his constitutional rights against search and seizure in any formal manner, as would be necessary for a search without warrant to be conducted legally. He testified at the hearing that he had been unaware that the police, who were his friends and who had kindly offered to bring him his car, meant to use it to construct sufficient evidence against him to charge him with the crime—that, in short, he had been tricked.

In hindsight, it is surprising that Joe gave the police his keys at all. Perhaps he didn't realize what it was that he had given them the chance to see. Two insignificant little bits of paper—he had possibly forgotten he even had them. And indeed, they didn't mean much until the police followed up the trail they indicated.

One of them was a receipt from the Holiday Inn at Yuma, Arizona, dated August 12, five days before the murder. This wouldn't have been unusual in itself, since Yuma was on the way from California to New Mexico. But the receipt was not made out in Joe Sneed's name. The name on the receipt was "Robert Crosset." The second slip was a receipt from a Surplus City store in Las Cruces, the town where Joe Sneed claimed to have spent the day and the night of August 17, when his parents were murdered. The date on the slip was August 17, and the nature of the purchase was not specified, but it gave the police the idea of inquiring at the store itself to try to discover what it had been.

When questioned about these papers by Captain Joe Barrios of the Silver City police, Sneed denied ever having used the name Robert Crosset. This raised the suspicions of the police even more. This suspicion connected with the use of a false name was what led directly to Sneed's arrest. By that time, the whole of Silver City was in a state of feverish excitement over the murder of two upstanding citizens.

The first act of J. Wayne Woodbury, Sneed's public defender, was to have his client file two motions to the district court of New Mexico. In the first motion Sneed requested a change of venue for the trial, on the grounds that the murder had "created a vast amount of publicity in the County of Grant, State of New Mexico, and much public excitement and local prejudice . . . together with misleading and erroneous stories carried by the Daily Press . . . [so] that a fair and impartial jury cannot be had." The elder Sneeds had both worked for the *Daily Press,* the local paper, and naturally it was covering their murder in lurid detail. Given the circumstances, the trial was moved from Silver City to Las Cruces in Dona Ana County.

In his second motion, in what must have appeared something of a life-or-death gamble, Sneed requested the suppression from the trial of the two receipts found in his car, asserting that at no time did he ever "give any representative of the Silver City Police Department any permission whatever to search his automobile." The motion to suppress evidence was accompanied by a superbly redacted brief citing a vast number of precedents in which the results of unlawful search and seizure were excluded from trial. The brief ends with a quotation from a celebrated judgment by the Supreme Court of the United States, a cry against the abuse by lawmakers and law enforcers of the very laws meant to protect citizens:

> The criminal goes free, if he must, but it is the law that sets him free. Nothing can destroy a government more quickly than its failure to observe its own laws, or worse, its disregard of the character of its own existence. Our Government is the potent, the omnipresent teacher. For good or for ill, it teaches the whole people by its example. . . . If the Government becomes a lawbreaker, it breeds contempt for law; it invites every man to become a law unto himself; it invites anarchy.

In spite of these efforts, Sneed lost his bet, and his motion to suppress the evidence was rejected. It was ruled that "an analysis of the evidence discloses no case of mere acquiescence, nor of mere submission to a demand, or to a show of force. An analysis of the law and the facts shows that valid consent to a search *can* be had—that in such cases no arrest, no search warrant, is necessary—and that this is the case at bar."

For E. C. (David) Serna, the district attorney prosecuting Sneed, it was particularly important to be able to present those two documents at the trial. This was because they tied in with a couple of further pieces of information that the police had been able to dig up once they had been furnished with the key name Robert Crosset. One of these was another record of a Robert Crosset, this time staying at a motel in Seaside, California, shortly before the sojourn in Yuma. The second and far more damning clue was that Robert Crosset was also the name of a person, described in the register as "a 5'9" male with brown eyes and hair," who had signed the sales register in a Las Cruces pawnshop upon purchasing a .22 caliber pistol on that fatal August 17. The buyer.had given as his address a post office box number in Las Cruces that was the same as the post office box number given by the Robert Crosset who had signed the hotel register at the Holiday Inn in Yuma. The only connection between this unknown Robert Crosset and Joe Sneed was that the receipt from Yuma had been found in Sneed's car. But that one connection suddenly made the case look black, very black, against Joe Sneed.

THE TRIAL began on February 1, 1965. In the prosecution's opening statement, they declared that they would prove the following facts:

- Sneed used the alias of Robert Crosset at two motels, in Seaside, California, and in Yuma, Arizona, and used the name again to buy a "cheap" .22 caliber pistol in a Las Cruces pawnshop.
- Sneed bought ammunition at a Las Cruces discount house, and a sales slip from the firm bearing the same date as the gun purchase was found in his car.
- Sneed's parents were killed in their bed by a .22 caliber pistol.
- There were no signs of forcible entry into their home.
- Sneed had a key to his parents' home.
- Sneed purchased a flashlight and gloves shortly before the slayings.
- Sneed's car was not parked at a Las Cruces motel on the night of the slaying as Sneed had told officers that it was.

Joe Sneed, "average" youth

If the prosecution could have proven all of these points beyond a doubt, Sneed's conviction would not have merited even a moment's hesitation. The trouble, however, was that it was not as easy as they made it sound to justify their claims. As the twenty-three witnesses for the prosecution underwent cross-examination one by one, the testimony about whether Sneed's car was or was not parked in Las Cruces on the night of the seventeenth turned out to be inconclusive—which was just as unfortunate for the defense as for the prosecution, since the whole defense strategy was based on establishing that Sneed had not left Las Cruces until the following morning. Similarly, the purchase of a flashlight and gloves was difficult to establish with certainty, as was the fact that the purchase Sneed actually made at Surplus City was ammunition, a fact that he denied.

The most difficult task of all for the prosecution was to demonstrate that Joe Sneed and Robert Crosset were one and the same individual. Certainly the hotel receipt from Yuma was found in Sneed's car, but there could be any number of explanations for that; Crosset could have been a hitchhiker or a hired killer, or Sneed may simply have picked up the receipt from the check-in desk in the hotel or from the ground of the parking lot, thinking it was his. To identify Robert Crosset with Joe Sneed—to condemn him, in effect, for murder—required far more solid proof. There were only three possible ways to obtain such proof, because there were only three people who could provide positive identification of Sneed as Crosset: the motel receptionist in Seaside, the hotel receptionist in Yuma, and the pawnshop salesman who had sold the gun. But at this point in the trial the prosecutors found themselves in a bind, be-

Edward Thorp, blackjack genius

cause as it happened, not one of the three was able to identify Joe Sneed for certain. The "average American youth" of average height, coloring, and features simply did not leave a strong impression in the memories of the people whose path he crossed.

Sneed's lawyer, Woodbury, put up a very simple defense; in substance, "We shall prove that this boy was not in Silver City the night of the murder." One by one, he destabilized the witnesses and weakened the weight of their evidence against Sneed.

At this point, the outcome of the trial was unclear. If the prosecution could convince the jury that Sneed and Crosset were one and the same, conviction was obvious. If they could not do so beyond a reasonable doubt, acquittal was a distinct possibility.

It was for exactly this reason that the prosecution finally chose to make a rather risky move, in the hope of gaining an edge. They called an unexpected expert witness: Dr. Edward O. Thorp, a young mathematics professor at New Mexico State University in Las Cruces who had recently achieved fame through the immense success of his book *Beat the Dealer,* which introduced card-counting methods that enabled enterprising gamblers to beat the casinos at blackjack. A man with a taste for adventure* and something of a local star, Thorp and his testimony were awaited with excitement by judge, jury, and spectators alike.

Dr. Thorp was the brother-in-law of prosecutor Ray Sinetar, who just two months before the Sneed trial had obtained a striking conviction in the Collins case (see chapter 2) with his new "mathematical proof of guilt." This was no coincidence: E. C. Serna of the prosecution had read the *Times* article about Ray Sinetar's mathematical proof and, struck by the possibility of using the same kind of argument in his own case, had telephoned Sinetar to ask if by any chance he knew someone who could play the role of mathematical expert witness at the Sneed trial. Working as he did in Las Cruces, Sinetar's brother-in-law Ed Thorp seemed like the perfect answer.

When Thorp received the unexpected phone call, he was intrigued, and agreed to have a talk with Serna for the purpose of learning more about the case, and seeing exactly what Serna wanted of him. Serna presented the case to Thorp in a way that made it look as though Sneed's conviction was such a foregone conclusion that Thorp could hardly understand the need for a probabilistic argument. Indeed, comparing what Thorp remembers being told† with the evidence that was actually presented at the trial, it seems that Serna was taking some care to let the mathematician feel that he was only being asked to use probability to confirm a guilt that was already clearly indicated by the evidence. Serna told Thorp that there was a motive: Sneed believed that his parents' interference had caused the breakup of a romantic relationship. He also told Thorp that it was known that Sneed had been in Yuma and in Seaside, whereas in reality this was not an established fact, but a deduction from the circumstance that Robert Crosset had been in those places, that the Robert Crosset receipt was in Sneed's car, and that the police believed Sneed to be Crosset.

While these elements are certainly convincing, it is not at all clear that they constitute proof beyond a reasonable doubt, which is precisely the rea-

---

*Such as, for example, donning a false beard and wraparound glasses to play blackjack in casinos where he was already known as the author of the book on how best to beat them. Thorp was well-known as a regular visitor in Las Vegas.

†Most of the information about E. Thorp's experience recounted here comes from a personal conversation with him over the telephone on September 23, 2012.

son for which Serna wanted the mathematician's testimony. Furthermore, the prosecutor told Thorp that Sneed had pointed out to police the exact place where he claimed that his car had been parked in a Las Cruces motel during the whole night of the murder, but that police had proven that Sneed's statement was a lie, as that particular parking place was located in an area that had been reserved that night for a firemen's convention. However, as mentioned above, the witnesses who testified about Sneed's parking place at the trial did not succeed in actually providing any incontrovertible proof that Sneed's car had or had not been there.

Finally, Serna explained to Thorp exactly what he wanted him to do: to ratify a probability calculation that he, Serna, meant to do in court, in imitation of what Sinetar had done in the Collins case. As Serna began to explain the arguments he proposed to use, Thorp perceived their defects and attempted to warn him, but Serna was clear about what he wanted to do. Convinced from all he had heard that Sneed's guilt was certain, and "always amenable to an interesting experience," Thorp agreed to play the game. In principle, he had nothing more to do than confirm the product rule on the stand, and multiply some probability figures together.

What Serna undertook to give the jury was a mathematical proof that the Robert Crosset of Yuma, Arizona (presumed but not proven to be Sneed himself), and the Robert Crosset who purchased a pistol at the Las Cruces pawnshop were one and the same person.

To begin with, he decided to calculate the probability that a random person might share a large number of traits with the Robert Crosset who had bought the gun. These traits included height, hair and eye color, post office box number, and of course the name itself.

The post office box number that had been given as an address in the pawnshop was the same as the one given in Yuma, and Serna observed that the chance of this happening purely by chance was 1 in 1,000.

The information about height and coloring obtained from the firearms sales register at the pawnshop was then presented in court. By law, when any weapon was sold, a description of the buyer had to be taken down, as well as the person's name and address. Robert Crosset was listed as being about 5'9" with brown hair and brown eyes. To calculate the probability of a man having these features, Serna handed the pawnshop register over to Thorp on the witness stand and requested him to examine and count the heights of purchasers listed there. Out of 35 purchasers, Thorp counted 12 having a height between 5'8" and 5'10" and 12 who had brown hair and eyes, concluding that each of the two types occurred with a probability of 12/35.

Finally, in order to calculate the frequency of the name Robert Crosset, Serna had several telephone books from various communities in the area brought in, and even invited the defense to bring in their own telephone books, which they did. The trial proceedings then came to a standstill as the judge, the jury, Serna, Thorp, Woodbury, and even Joe Sneed himself all examined the phone books to see if they could find any Crossets. (Sneed's apparent indifference to the situation as he coolly examined the phone books "as though it was a school exercise" did nothing to convince spectators of his innocence.)

From this counting, Serna estimated that about 1 in 30 men bore the first name of Robert. But the last name Crosset did not occur in any of the phone books examined in the courtroom that day. Estimating that the books contained a total of about 1,290,000 names in total, Serna guessed the frequency of the name of Crosset in the general population to be about 1 in a million.

He then asked Thorp on the stand about the use of the product rule in this situation. Thorp explained that, assuming that Serna's probabilities were correct and that the events referred to were independent, the probability of them all occurring together would be their product, and he multiplied the numbers 1/30 (for the name Robert), 1/1,000,000 (for the name Crosset), 1/1,000 (for the post office box), 12/35 (for the height), and 12/35 (for the coloring). According to this calculation, the probability that a random person in the United States would have the same name, height, hair and eye coloring, and post office box as the man in the pawnshop came out to roughly 1 in 240 billion. "The significance of this figure," Thorp is quoted as having stated on the stand, "is a chance of 240 billion to one that it was our suspect who was responsible for this series of numbers in the pawnshop, as opposed to some person coming in and accidentally implicating him. This is the application of this thing to criminalistics."*

In cross-examination, Sneed's lawyer, Woodbury, did put his finger on the problem of estimating the probability of an "extremely rare" event, if in

---

*This direct quote comes from Judge Wood's opinion at Sneed's appeal trial. Thorp does not remember making such an explicit statement, and in particular finds it unlikely that he ever would have employed a word like "criminalistics."

a somewhat unexpected manner. "What is the probability of the waters of the Nile running red with blood?" he asked the surprised professor of mathematics. Suspecting that the lawyer was trying to lead him into rebutting a biblical fact in front of a jury from deeply religious Las Cruces, Thorp eluded the difficulty by responding: "Assuming that event occurred, it was only once in the known history of mankind, and a probability figure cannot be associated with it."

The point Woodbury was trying to make is a very good one, but he probably did not have the mathematical savvy to demolish the numerical testimony by following it through, or by citing Math Error Numbers 1 and 2: the inaccuracy of the probability figures and the non-independence of the events they described. The jury may not have made the connection between the waters of the Nile and the name Crosset, or they may have trusted a mathematics professor to know his figures better than a Bible-quoting lawyer. In any case, the prosecution's arguments convinced the ten-man, two-woman jury, who after a deliberation of seven and a half hours declared the defendant guilty of murder in the first degree, with a recommendation of life imprisonment.

Upon being asked if he had anything to say before the judgment and sentence of the court were passed upon him, Sneed stated that he was not guilty. Still claiming innocence, still denying that he had ever used the name Robert Crosset, Sneed was transported directly from the court to the New Mexico State Penitentiary in Santa Fe to begin serving his life sentence.

NO SOONER had Sneed settled into his new residence than he sat down with his lawyer and wrote down all the grounds on which he might appeal his conviction. "The defendant believes that the District Court made grave and reversible errors in the trial of this cause, which should be reviewed by the Supreme Court in the State of New Mexico," Sneed and Woodbury wrote in filing a motion requesting a free transcript of the trial testimony, as the prisoner was "destitute and without funds." The motion was granted, and by July the men were in possession of all the documents they needed to prepare a beautifully written, well-thought-out, impeccably argued appeal, which, although they did not know it at the time, would become a historic document.

They gave three grounds for appeal: unlawful search and seizure, improper comment to the jury concerning Sneed's choice not to testify, and, most important by far, the erroneous use of mathematical probability to establish an otherwise unproven identification.

Basically, the mathematical part of the argument given by the prosecution comes down to the following: the probability of two completely different people giving the same name (Robert Crosset) and the same post office box number (210) and having the same height and hair and eye color, purely by chance, was computed as just 1 in 240 billion, therefore they must have been the same person, and from the evidence of the receipt in the car, that person must be Joe Sneed.

There are many shaky features in this approach. One of the weak points is the final deduction. It may not seem very useful to identify the two Robert Crossets with each other, when the identification of the Yuma Crosset with Sneed was not proven, nor was it even known that Sneed had been in Yuma at all. Although Serna had tried to imply that the presence of the receipt in Sneed's car established his identification with the Yuma Crosset, neither he nor the defense ever proposed, for example, that it could have been placed there by someone framing Sneed. The presence of the receipt in the car made the identification of Crosset with Sneed appear very likely, but certainly not beyond a reasonable doubt.

Then, there are all the problems with the actual probability calculation itself. By estimating the frequency of this name in the general population, Serna appears to have been assuming that the name was a real rather than an invented one. But at the same time, Serna was trying to prove that Robert Crosset was Sneed, in which case the name Robert Crosset would have been invented, so that any calculation concerning the frequency of the name in the actual population would not be relevant; indeed, there is no reason to suppose that the frequency of an invented name, for example, John Doe, corresponds to the real frequency of John Does in the population.

And finally, the details of the calculation itself do not hold up. The probability of 1 in 1,000 that two people would randomly choose the same post office box number is valid only if all post offices have exactly 1,000 boxes. Actually, that probability diminishes when post offices are larger and have more boxes and increases when they are smaller.

Next, it is not reasonable to estimate the proportion of the male population of the United States measuring between 5'8" and 5'10", or the proportion with brown hair or brown eyes, from one list of thirty-five gun purchasers in one pawnshop in one particular town, whose ethnic population distribution may be very different from that in other parts of the country. Such a small sample is almost certain to lead to very inaccurate estimations.

Then, it is not legitimate to multiply the probabilities of having the right height, hair color, and eye color. Indeed, height and coloring are certainly not independent features; in a city with a large Hispanic population, for instance, a shorter stature will often be associated with black hair and eyes.

Finally, by citing the arbitrary figure of 1 in a million for the estimated frequency of the name Crosset in the population of the southwestern United States, the prosecution missed the possibility that there could be a whole group of Crossets—say, belonging to one extended family—in one specific location for which they had not consulted the phone book. The absence of a name from a set of phone books actually gives very little information about its frequency in a region of the United States, because in trying to use phone books to estimate the frequency of the name occurring in the country, one is actually working under the unjustified assumption that the name is evenly distributed around the country.* One simply cannot draw any conclusion more precise than to note that the name is not very common. Zero cannot be given a positive measurement.

Then, the multiplication of the probability concerning the first name Robert with the estimated probability for the last name Crosset is also wrong, since again, these two events may not be independent. Indeed, there is no reason for the distribution of the first name Robert among all Crossets to be equal to that in the general population, since it often happens that in an extended family, the same first name occurs with much greater frequency due to family tradition.

In summary, the computation of a probability for two Robert Crossets with similar features not to be the same person is essentially meaningless in the context of Sneed's trial. The number may sound convincing, but it simply does not correspond to any real measure of anything. Yet that calculation, that 1 in 240 billion figure, convinced a jury that Sneed was Crosset, and on the strength of this erroneously determined fact a man was condemned for

---

*For instance, the name Schneps is quite rare, and whenever two people of that name meet, if they are able to trace their ancestry back three or four generations, they generally find that they have common relations all coming from a single town in Galicia (now Poland).

murder. It was only on appeal that the prosecution's reasoning was revealed to have caused more confusion than clarity.

On May 31, 1966, just over a year after the original sentence, the supreme court of New Mexico annulled the previous verdict and ordered a new trial for Sneed. The summary of the judgment shows that while the justices understood that there had been problems with the mathematical reasoning introduced in court with Thorp as expert witness, they were uncertain about the correct course of action.

> [Thorp did not] state why a positive number was used in arriving at an estimate on the basis of the telephone books when the name Robert Crosset was not listed in those books. Since the name Robert Crosset did not appear, should any estimate, based on the telephone books, be used at all? Or should a zero be used as the estimate based on the telephone books?

Sneed's retrial was to take place without any recourse to mathematics.

THE REST of the Sneed case is worth recounting because it shows just how difficult it was, in the absence of identifying information and without recourse to doubtful mathematical reasoning, for the prosecution to prove its case. Mathematics or no mathematics, the investigators were convinced that Sneed was guilty, and so for a second time the prosecution set out to see what evidence it could provide to convince the jury that Sneed should be identified with Robert Crosset; Sneed himself continued to deny having ever used the name. The new trial began on August 16, 1966.

The prosecution began by presenting the receipt from the Surplus City shop in Las Cruces that had been found in Sneed's car. Although the nature of the purchase was not recorded on the receipt, they were able to confirm from the shop's sales records that standard ammunition had been purchased in the shop on that day. They also presented the witnesses who declared that on the day of the murder Sneed's car could not have been parked in the lot of the Las Cruces motel where he claimed to have spent the night, due to the firemen's convention. The witnesses' testimonies turned out to be rather shaky on cross-examination, since a person can and often does park his car in a reserved place, and it was impossible to actually prove that

Sneed had not done so. At the same time, though, Sneed could not bring forward a single witness to support his alibi, nor did he give any explanation for the Crosset receipt in his car. In fact, the appeal trial had reached the exact same point where, in the original trial, E. C. Serna had decided to apply a probabilistic proof of identity. But now that avenue was no longer open. Another solution was needed, for the police investigators were certain that Sneed had killed his parents, and that an acquittal would simply let a cold-blooded murderer go free.

No one knows exactly what methods the prosecution used to get their main witnesses to change their statements, but when the clerks from the Yuma and Seaside hotels took the stand, their testimony was dramatically different from what it had been at the first trial. Clarke Wallace Fowler from the Holiday Inn in Yuma stated under oath that Joe Sneed was the man who had registered there on August 12, 1964. He claimed to recognize him and pointed his finger squarely at the suspect sitting in the courtroom. Marilyn Moore, the receptionist from the motel in Seaside, followed Fowler on the stand and did the same. On the basis of these identifications of Sneed with Robert Crosset, the prosecution rested its case.

But Sneed's defense attorney was aware that the two clerks had not been able to identify Sneed at the first trial, and he had no intention of letting them get away with doing so now. Under cross-examination Fowler was forced to admit that he had been unable to identify Sneed at the preliminary hearing in 1964. "How is it that you failed to recognize him then, barely weeks after the murder, and yet now you are so certain?" asked Woodbury. Stammering, the witness replied that he would have been able to identify him the first time, except that he had been "too nervous." Woodbury said he had no further questions. The power of Fowler's testimony was shattered. Cross-examining Marilyn Moore, Woodbury asked her the same question: how could she be so sure that the young man on trial was the same one who had stayed at her little road motel two years earlier? "I know it was him," she replied, "because he appeared neat and clean, and that is outstanding for our area." The spectators laughed.

The case of Joe Sneed might have ended like that of Lizzie Borden, with an acquittal due to lack of evidence even though everything seemed to point toward guilt. But the prosecution had one more string to its bow: a last-minute witness whose testimony had nothing to do with "Robert Crosset" or Sneed's trip from California to New Mexico, or even the murder of Joe and Ella Sneed. Instead it provided a window into the soul of the young man who denied the charges against him and sat, silent and inscrutable, at the defense table.

What the investigators discovered was that in spite of his young age, Joe Sneed had gotten married while in California, but the couple had divorced. According to information given by the prosecution to Edward Thorp outside of the courtroom, Sneed's motive for the murder of his parents was their interference in his relationship with the woman, but no evidence of this had been presented at the first trial. The young woman was now remarried to someone else, but she agreed to testify as a witness at Sneed's second trial. Her name was Kathy Storey, and the tale she told clinched the case for the prosecution as much as any identification ever could have done. The domestic violence she described soon made its way into the press: she vividly recounted how her husband had hit her in the head with a book and bitten her in the face, and how he had once thrown a fountain pen at her with such force that the point had remained embedded in the flesh of her leg. She also told a story of how he had once picked up the family dog and, before her horrified eyes, slammed the defenseless creature into the wall with all his strength.

Mrs. Storey was about to start speaking of things that Sneed had said to her during their short marriage—in particular about his parents—but Wayne Woodbury cut her short. Her testimony would be hearsay, he declared, and if she were allowed to give it, he would call a mistrial. The judge called a recess to think this over and consult the law, then agreed with Woodbury. He considered that Mrs. Storey's testimony was sufficient as it stood.

Woodbury had spent the five days of the trial trying (relatively successfully) to destroy testimony showing that Joe Sneed was Robert Crosset, and (relatively unsuccessfully) to prove that Sneed had been at the motel in Las Cruces, with his car parked in the parking lot, throughout the evening of the murder. Kathy Storey's testimony, coming as it did from such an unexpected angle, blew up his entire strategy. There wasn't much to say in response.

Woodbury called a single defense witness to the stand, Sneed himself, and asked him a single question.

"Did you shoot your parents?"

"No, sir. I did not," was the reply.

And Woodbury rested his case. There was nothing more to add, nothing else that he could do. As Grant County Assistant District Attorney William

Martin explained, in asking for the death penalty during a one-and-a-half-hour summation, "The 23 prosecution witnesses have proved during this five-day trial that Sneed lied for his life."

The eleven-man, one-woman jury went out at 6:10 p.m. and deliberated until 1:20 a.m. The verdict was guilty of murder in the first degree, with a recommendation of life imprisonment. Sneed was impassive as District Judge William Scoggin pronounced sentence. Asked by Scoggin if he had anything to say, Sneed answered, "Only what I said two years ago: I'm not guilty."

Willie Silva, the Dona Ana County deputy sheriff who accompanied Sneed directly after the sentencing, reported to the newspapers that "he didn't look disturbed at all. He didn't say a thing. He was quiet all the time. Not a goodbye, or a hello: nothing."

NO SOONER had Sneed been returned to his prison cell than he pleaded with his lawyer to lodge a second appeal and take the case all the way to the supreme court of New Mexico. He must have realized by then that without the two receipts found in his car during an unlawful search, there would have been virtually no evidence against him at all. Having the receipts suppressed was his only chance, but it was a significant one. Condemned to life in prison, Sneed was grasping at straws.

Ten days after the end of the second trial, his lawyer sent a letter to the judge, the tone of which speaks clearly enough of the utter hopelessness of the cause, and hints more than strongly that Sneed's own defense attorney was convinced of his guilt.

Dear Judge Scoggins,

Joe E. Sneed[,] who was recently convicted of first degree murder in Dona Ana Criminal Action No. 11232, is insisting that I appeal his conviction to the Supreme Court. Inasmuch as I have gone this far with Mr. Sneed, I do not feel that I can, at this time, forsake him. However, I would like to be able to acquire actual expenses in connection with the appeal. Also, I am wondering whether it will be necessary to file additional pauper's affidavits in order to obtain free process.

I will look forward to hearing from you in this regard.

With kind personal regards,

Sincerely yours,
J. Wayne Woodbury

On December 20, 1967, the supreme court of New Mexico affirmed the judgment and sentence, and Joe Sneed faded into history as another murderer, never fully explained, never fully understood. But the attempt to prove guilt by probability, and the state supreme court's overturning of that attempt, reached legal tentacles into the future whose importance far outstripped the significance of the case itself.

MATH ERROR NUMBER 4 »
DOUBLE EXPERIMENT

AS THE PREVIOUS CHAPTERS have shown, probability is a delicate subject, because it can often run contrary to elementary intuition. We saw that even if the probability of an event occurring is correctly determined, it can be wrong to multiply it by itself if independence of the occurrences is not guaranteed. In this chapter we consider another common error concerning multiple occurrences.

Suppose you are running a test with a yes or no answer—for example a diagnostic test for an illness—and suppose that a positive result indicates that the illness is really present with a probability of 60%. You run the test and get a positive, meaning you can be 60% sure that what you are testing for is effectively present. Is it worth running the test again? If you get another positive, does it simply again indicate that you can be 60% sure that what you are testing for is present?

As it happens, there is a benefit to retesting: the combined effect of the two experiments makes a much stronger case than the results of each one separately. Let us show exactly how this works by using a simple example.

Suppose you are given a coin and told that it is of one of two types: either fair and balanced or weighted to come up heads 70% of the time. You are allowed one toss, and it falls on heads. Let us first investigate the probability that the coin is biased after this result.

To determine whether the coin is fair or weighted, we need to calculate the probability A of falling on heads if it is a fair coin, the probability B of falling on heads if it is a biased coin, and then multiply both A and B by a scaling factor that will bring the total probability to 1. (We know that the coin must be either fair or biased—there are no other possibilities.) The scaling factor will therefore be C = 1 / (A + B), with the final probability of the coin being fair equal to A × C, and biased equal to B × C.

In our example, the probability that the coin will fall on heads is A = .5 if it is fair, and B = .7 if it is biased. Thus C = 1 / (A + B) = 0.8333 . . . and the coin has a probability of being fair equal to A × C, roughly .416, or 42%. The probability of the coin being biased is B × C, which comes out to about .583, or roughly 58%. So, since the outcome of the single-toss experiment was heads, the conclusion that can be drawn from that outcome is that the coin is biased with a probability of 58%.

Now suppose you do the same experiment a second time. You toss the coin and it again falls on heads. By the previous calculation, you have again found that the coin is biased with a probability of 58%. But what happens when you combine the two results, considering them as one double-toss experiment whose outcome is two heads rather than as two separate, independent experiments?

The procedure to calculate the probability of the coin being fair or biased after this double experiment is exactly the same as before: first we calculate the probability A of the outcome for a fair coin, then the probability B of the outcome for a biased coin; we let C = 1 / (A + B) be the scaling constant as before, and A × C and B × C give us the final probabilities that the coin that twice came up heads was fair or biased, respectively.

Coin tosses are independent, so we can multiply their probabilities. Thus, the probability of two consecutive heads coming up with a fair coin is A = .5 × .5 = .25, and the probability of two heads with the biased coin is B = .7 × .7 = .49. We find C = 1 / (A + B) = 1.3513, so the probability that the coin is fair is now A × C = .337, or about 34%, whereas the probability that it is biased is B × C = .662, or about 66%. Thus, getting the same result twice under the same conditions has increased the reliability of the result from 58% to 66%!

In the case we examine now, the judge made the error of assuming that a new DNA test on a presumed murder weapon would provide no more information than the first one, and chose to reject a second test that might have proved decisive.

The Case of Meredith Kercher: The Test That Wasn't Done

It was November 1, 2007. Meredith Kercher, a British student spending a year in the medieval Italian city of Perugia under the Erasmus program, spent a quiet afternoon having pizza with friends and watching a movie. She left her friends' apartment a little before 9:00 p.m. and a few minutes later reached her home, a pretty cottage just outside the city walls, which she shared with two Italian girls and an American girl named Amanda Knox.

Around the same time, Raffaele Sollecito, a young Italian student living on a busy street a short distance away from the isolated little cottage where Meredith rented a room, used his computer for the last time that evening. One week earlier, he had met Meredith's housemate Amanda at a classical

Meredith Kercher

Raffaele Sollecito

concert, during which a string ensemble played Schubert's Trout Quintet and tangos by Astor Piazzolla. The two of them had talked during intermission, spent the night together at Raffaele's place afterward, and become practically inseparable from that point on. Raffaele was a shy student who had previously never had a girlfriend; he favored computer science, violent manga, and knives. Amanda, an outdoorsy type, didn't seem to mind his introverted nature, appreciating her new lover's devoted tenderness.

On the evening of November 1, Amanda Knox received a text message from her boss informing her that she didn't need to come to work that night, because business was slow. Le Chic was a trendy pub in the center of Perugia run by Congolese musician Patrick Lumumba, and Amanda had found a job waitressing there a couple of nights a week. But according to the testimony she gave at her subsequent trial, when she found out that she didn't need to go in to work, she spent the entire evening at Raffaele's place, watching a film, eating dinner, smoking pot, making love—as she made sure to explain to an attentive jury—and sleeping.

On that same evening, Meredith was stabbed to death in the little hillside cottage where she and Amanda both lived. The results of the autopsy revealed that she had been attacked by more than one person. Indeed, the knife wounds she received to the neck were made not only from different angles—one from the right, one from the left—but also by different knives, according to the size of the wounds; a bloody print on her bedsheet showed where a knife had been laid down briefly on the bed. Innumerable bruises and contusions on her body showed that she had been gripped, restrained from defending herself, abused, and choked before finally being killed.

In her courtroom testimony Amanda described how she returned to her cottage the following morning to collect some clean clothes and take a shower, but noticing a few "strange things," she became worried and went back to Raffaele's place to tell him about them. She brought him over to the cottage and showed him the things that had unsettled her: some small traces of blood in one of the bathrooms, unflushed excrement in the other, and Meredith's bedroom door locked. They explored the house together and discovered something worse: the window in the bedroom belonging to Filomena, one of the Italian girls, was smashed, and the room itself had been ransacked. Amanda called Filomena and told her to come home at once.

Filomena came, bringing several friends, and the police came as well—not the Carabinieri initially, but the Italian Postal Police, who had been independently investigating the discovery of Meredith's two cell phones in a nearby garden, where they had been tossed and then found by an elderly woman who lived there. Filomena panicked at the sight of her ransacked room and Meredith's locked door and insisted on having it broken down. Her friends kicked it in, and there lay Meredith's body on the floor in a welter of dried blood, covered with the quilt from her own bed.

It is not clear just when Amanda and Raffaele stopped being witnesses in the eyes of the investigators and began to be considered as suspects. At any rate, it is certain that even from the start there were a number of little details that alerted police to something peculiar. The smashed windowpane and the raid on Filomena's room looked fake: nothing was missing, and there were no footprints, trampled grass, marks on the wall, or shards of fallen

The "house of horrors," as the pretty cottage was called in the press

glass on the ground below the broken window outside the house, nothing at all to confirm the hypothesis that someone had really made the difficult climb from the outside. The investigators reasoned that if anyone had staged a break-in, it had to have been someone eager to make the murder look like an outside job. But this would only be necessary if the person was actually an insider—even an inhabitant of the house.

Worse, when questioned, Raffaele first stated that he and Amanda had been at a party with friends on the evening of November 1, but when said friends could not be produced, he then recalled that they had spent the evening at home.

The growing suspicions of the police may or may not have been rein-forced by Amanda's odd behavior in the days following the murder, during which she flirted with her boyfriend at the police station in front of Mere-dith's grieving friends and responded to questions about the murder with such flippant remarks as "Shit happens" and "She fucking bled to death." Indeed, later on much would be made of the idea that she was arrested be-cause of her unbecoming behavior. But there was certainly more to it than that. The pair was questioned both together and separately multiple times over the following days, and late on the evening of November 5 Raffaele suddenly announced that he had told lies for Amanda and that while he had stayed at home surfing the Internet—a lie, as was subsequently shown by the records of his computer activity—Amanda had gone out alone. At least she might have. He couldn't remember. He was too stoned.*

On hearing all of this, the police took Amanda, who had not been sum-moned to the police station that evening and was cheerfully doing homework

---

*The final version of events presented in court by Raffaele's lawyers (he himself did not tes-tify), and in his recently published book, *Honor Bound,* is that they remained together at home throughout the evening. But it contrasts not only with his declarations to police in the days following the murder, in which he said that she had gone out alone, but also with his written statements in his diary from prison, in which he said: "I remember that . . . Amanda had to go to the pub where she usually works, but I do not remember how long she was gone. I remember that she subsequently told me that the pub was closed (I have serious doubts regarding the fact that she went out). I am straining myself to remember other details but they are all confused." Finally, cell phone ping analysis indicates that Amanda was in Raf-faele's apartment when the text message from Patrick arrived in her phone at 8:18 p.m., but a few streets away when she actually answered it, which did not happen until 8:35 p.m., al-most twenty minutes later.

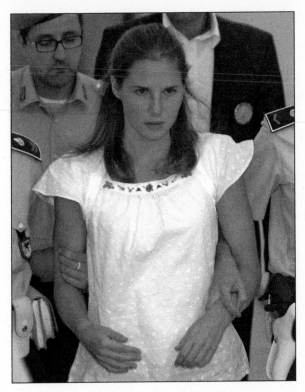

Amanda Knox being brought to court

and gymnastics in the hall, into a separate room and questioned her intensively. The focus of their questioning was precisely whether or not she had gone out on the evening of the murder. In her phone they found her text message to her boss, acknowledging his (deleted) message about not coming in that evening. She had written, "Ci vediamo piu tardi. Buona serata," thinking this was the normal Italian for "see you later," but in fact it means something closer to "see you later today" (or "see you tonight"); in other words, it indicates a planned meeting. The police seized on this message and insisted on knowing to whom it had been written. Amanda says they shouted at her and even cuffed her on the back of the head, though the police deny this.

What happened next was a bombshell: Amanda broke down in tears, suddenly declaring that Patrick Lumumba was the murderer and that she had been in the cottage while he did it, cowering in the kitchen and covering her ears to block out the sound of Meredith's screaming. She retracted this confession the next morning in a bizarre written statement in which she explained that the memories in her mind were "blurred flashes" and seemed less real than her memory of having been with Raffaele. Later she would claim that the story was all an illusion provoked by the intense pressure put on her by the police and their pressing suggestions that she had seen the murder and suppressed the memory. Nevertheless, her confession led to the arrest of Patrick Lumumba, Raffaele Sollecito, and Amanda herself, accused of being present at the scene. The police were also searching for a fourth man, whose yet unidentified footprints and DNA traces had been found in various places around the house, on Meredith's clothing, and inside her body.

AFTER RAFFAELE'S arrest the police made a search of his apartment, and from his cutlery drawer they took a large kitchen knife. The investigator who collected it stated that it looked and smelled particularly clean and was lying on top of the other, ordinary table knives. This "very clean" knife was delivered to the laboratory of the Scientific Police of Rome for forensic analysis and handed over to their forensic geneticist, Dr. Patrizia Stefanoni.

According to her testimony, Dr. Stefanoni began by unpacking the knife from its wrappings and examining it under a strong light. She perceived a few streaks on the blade, perhaps indicative of vigorous scrubbing. She took swabs along these streaks, managed to obtain an infinitesimal quantity of biological material—human cells, in fact, but precious few of them—and proceeded to perform two analyses on it: one to determine whether or not the cells were blood cells, and the other to obtain a DNA identification. She also swabbed the handle of the knife.

The DNA on the handle turned out to be Amanda's, but this was not considered incriminating, as she had used the knife while cooking at Raffaele's house. The DNA on the blade, however, was a different story. The first test could not establish whether or not the cells were from blood. They were human cells, but they might have been from skin or tissue.

For the second test, Stefanoni's machine was set up to work with a minimal quantity of biological material that was significantly greater than what she had at her disposal. At first this machine simply output "too low," indicating that using the standard settings it was not able to analyze such a tiny quantity. Many forensic geneticists would have followed the accepted rules and stopped the analysis at this point. But Stefanoni elected to continue, modifying the settings and pushing the machine beyond the limits recommended in the manufacturer's guidelines. By doing this she managed to obtain a DNA profile for the cells on the knife.

A DNA profile is given as a graph called an "electropherogram," which shows a set of "genetic loci": pairs of peaks situated at particular locations along a horizontal line. Every human being possesses millions of gene pairs, each of which has been given a name, but there are thirteen particular gene pairs that have been singled out by geneticists because the pairs differ quite significantly from person to person. The probability that two people (excluding identical twins) could have all of these thirteen peak pairs at the same locations in the electropherogram has been estimated at around 1 in 400 trillion, many times the population of the earth. To be certain that the graphs from two samples both come from the same person, every single peak must lie in the exact same position. If even one peak is clearly in a different location, then the two samples necessarily come from different people.

The figure below shows an electropherogram of Meredith Kercher's DNA taken from a swab. The thirteen pairs of peaks are clearly visible along a horizontal axis (which has been broken into three lines to fit on the page). The vertical axis shows the heights of the peaks, which are measured in units called RFU (relative fluorescent units). When the DNA sample is sufficiently large (such as the one that produced the graph below), peak heights tend to reach as high as 1,000 or 2,000 RFU.

This electropherogram was created in the Scientific Police laboratory as a reference to determine whether other DNA samples from the crime scene came from Meredith. It is of good quality, with high, clearly defined peaks that cannot possibly be confused with the normal tiny peaks from background noise (sometimes called "stutter") that appear all along the axis. If all DNA samples gave such clear results, DNA analysis would be a more precise science.

But the fact is that subtleties can be involved in DNA analysis, particularly in cases where the

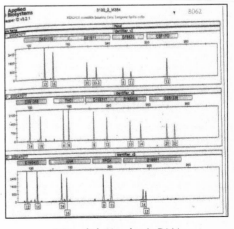

Meredith Kercher's DNA

DNA sample is degraded, contains a mixture of DNA from more than one person, or is extremely small. The graph of a degraded sample may show only a few peak pairs out of the usual thirteen; a mixed sample will show too many peaks that are difficult to pair correctly. When the DNA sample is particularly small—such samples are called LCN, for "low copy number"—the heights of the peaks in the electropherogram are correspondingly much lower than the 1,000 or 2,000 RFU that will appear in a good sample such as the one above. Forensic geneticists are trained in methods to distinguish small true peaks from the occasional extra-large peak from background stutter in the output from LCN samples, but it can be difficult to do so with certainty, and experts may not always agree with each other in interpretation. One commonly used guideline is that any peak less than 50 RFU in height is subject to doubt.

In the case of the cells found on the knife seized in Raffaele's flat, the problem was precisely this very small sample size, which Stefanoni attributed to the knife's having been washed; indeed, the few cells that did remain were lodged inside a scratch in the metal. For this reason, Stefanoni was not able to apply the first most basic technique of DNA analysis—namely, dividing the sample into at least two smaller parts so as to compare the graphs from two independent runs through the machine. With two graphs, there is a very simple method for determining true peaks, which is to accept only those that appear in the exact same place in both graphs. This is considered valid even for small samples that produce low peaks, since there is virtually no probability that the random background noise would produce an unusually large peak twice in the same position. But Stefanoni feared that if she divided the already minuscule sample into two, she would obtain no results at all. She took the chance of using the entire sample on a single run.

The figure on the left shows the electropherogram from Stefanoni's analysis of the DNA from the cells on the knife. All thirteen pairs of peaks are clearly visible,

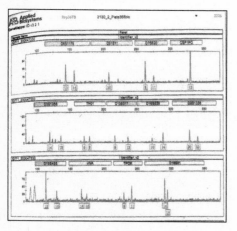

The knife-blade DNA

and the background noise is extremely low except for a few extra peaks that are about the same height as the visible pairs. They are all very low compared to the peaks from the abundant sample of Meredith's DNA; the vertical axis of the graph shows that many of the peaks in the knife sample are less than 50 RFU, or below the accepted minimum. However, it is important to understand that that number serves as a guideline to distinguish low peaks from high stutter. In some cases, the stutter remains minimal and the profile appears clearly despite the relatively low peak heights. That is the case here.

Stefanoni proceeded to compare this output with the graph from the known sample of Meredith's DNA, which we do below by superimposing them one on top of the other. The graphs have not been scaled for height, which is not connected to the identification of the individual. The only thing that counts for identification is the exact placement of the peaks along the axis. In these graphs, all the peak pairs in the two samples correspond

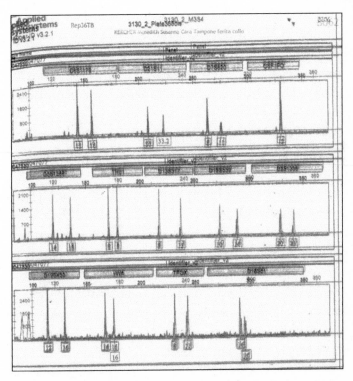

The two DNA samples: Meredith in thin, knife blade in thick

perfectly, and Stefanoni logically concluded that the DNA on the knife blade also belonged to Meredith. To assume that this DNA is not Meredith's would be tantamount to asserting that background noise randomly produced some unusually high peaks in the precise places where Meredith's DNA peaks would normally be, a probability so low as to be negligible.

WHILE STEFANONI was performing her analyses in the lab, police investigators were making progress on the outside. On November 12 police found a witness—the only client at Le Chic that night—who asserted that Congolese pub owner Patrick Lumumba had spent the entire evening at his bar, providing him with an alibi. Patrick remained in prison while the police worked on confirmation.

On November 13 it emerged that police were still searching for the "fourth man," the one whose DNA had been detected in the bathroom, on Meredith's handbag, and on her body. This DNA had been analyzed and proved to belong to neither Amanda, nor Raffaele, nor Patrick.

On November 15 Dr. Stefanoni dropped her bombshell: the knife DNA was Meredith's.

On November 16 it was leaked that the identity of the fourth man was known, but that he had fled from Perugia.

On November 18 it was revealed that the fourth man was African.

On November 19 the fourth man's name emerged: Rudy Hermann Guede.

On November 20 Guede was arrested in Germany, where he was found train hopping without a ticket. On that same day, Lumumba was released.

On November 25 Guede, still held in Coblenz, made a kind of confession to the German police. He told a strange tale of having been in the middle of fooling around with Meredith when he suddenly had an urgent need to visit the toilet, during which time an Italian stranger came in and murdered the British girl. Hearing her scream, Guede said, he rushed out of the bathroom so quickly that he had no time to flush, and he attacked the murderer, who was wearing a swimming cap and brandishing a knife in his left hand. But Guede's unfastened pants slid down to his ankles, causing him to fall over and giving the unidentified killer the chance to escape with a final shout of "Black man found, black man guilty!"

Guede was brought back to Italy on December 6 and promised that he would tell the full story the next day. After seven hours of interrogation, the same story stood, with the additional detail that there had been another person standing outside, who had fled together with the killer. Guede expressed both guilt and grief at having failed to save Meredith, who had been dying when he saw her. He told how he had fought with the killer and received some tiny cuts to the fingers, which still showed. He recounted how he had tried to staunch her bleeding wounds with towels from the bathroom (which were indeed found at the crime scene), and how he had written the letters "AF" on the wall in blood, thinking she had tried to pronounce this syllable with her dying breath (although no such letters were found). He had not called the police, he said, because he had no cell phone, and Meredith's had been stolen by her killers. Afraid of being accused of the murder, Guede had run away.

It seemed that in Rudy Guede the police had found the murderer they sought. A high-school dropout with a miserable family background, he lied easily, couldn't hold down a job, was well-known as a good source for students to get hold of a bit of hash, and had already been involved in a little petty burglary. He admitted to being in the house at the time of the crime. His traces were in the room and on Meredith. No one believed the story about the pair fooling around, as Meredith was a serious girl who had openly expressed her dislike of cheating and was seeing another boy at the time. Worse, it emerged that after leaving the grisly crime scene he had spent the rest of the night dancing at a local disco. And of course his fleeing the country didn't help his image. It seemed as though the guilty party had been identified and would soon be brought to justice. Amanda, Raffaele, and their families should have been dizzy with relief.

Except for one problem: the evidence against them wouldn't go away. Most incriminating was the DNA found on the knife from Raffaele's apartment, which Meredith had never visited. This knife, appearing on a television report of the crime investigation, was brought to the attention of both Raffaele and Amanda in their prison cells.

Raffaele sank himself deeper into trouble by writing peculiar things in his diary. When Rudy Guede was arrested, for example, he wrote: "Today finally they have captured the real murderer of this incredible story. He's a

22-year-old Ivory Coaster and they found him in Germany. Papa was happy and smiling, but I am still not 100% calm, because I'm afraid that he will invent strange things." One really wonders what "strange things" he was expecting to hear from the mouth of that supposedly unknown stranger.

Even more oddly, when he learned that Meredith's DNA had been found on his own kitchen knife, he wrote: "The fact that Meredith's DNA is on the kitchen knife is because once while we were cooking together, I moved around holding the knife and pricked her hand. I apologized right away but she wasn't hurt. So the real explanation of the kitchen knife is this." It was easily proven that Meredith had never been to Raffaele's place. Furthermore, her DNA was not on the tip of the knife, as would be expected from his story, but in a scratch on the flat of the blade. And he had never mentioned such an episode before during all the hours of interrogation; it came out only when police suddenly found a speck of Meredith's blood there. The situation didn't look good for Raffaele.*

In her own diary Amanda wrote that it was impossible that the DNA on the knife could be Meredith's, since Meredith had never been at Raffaele's house. She then slid into vague musings about whether it was possible that Raffaele had taken the knife and slipped out to murder Meredith, then returned and pressed it into Amanda's hand while she slept in his bed.

Rudy or no Rudy, the knife began to look devastating for the lovebirds.

There was only one hope. Dr. Stefanoni had made use of exceptional methods to test the knife DNA. Because there were so few cells, she had not been able to divide the sample into two parts, so the test she had made on that sample could not be repeated. For the defense effort spearheaded by Amanda's family, the best option now was to discredit the knife.

MERE WEEKS after the murder—in the very heat of the investigation—Amanda Knox's parents, Curt Knox and Edda Mellas, hired the Seattle-

---

*In his book, *Honor Bound*, Raffaele explains his explanation thus: "How did Meredith's DNA end up on my knife when she'd never visited my house? I was feeling so panicky I imagined for a moment that I had used the knife to cook lunch at via della Pergola [Meredith's home] and accidentally jabbed Meredith in the hand. Something like that had in fact happened in the week before the murder. My hand slipped and the knife I was using made contact with her skin for the briefest of moments. Meredith was not hurt, I apologized, and that was that. But of course I wasn't using my own knife at the time. There was no possible connection."

based communication strategies firm of Gogerty and Marriott to orchestrate a public relations campaign of enormous dimensions. By January 2008 Curt and Edda were in Perugia with an ABC television crew, traveling by limousine, staying at high-end hotels, and, of course, visiting their daughter. And that was only the beginning. Over the course of the following months and years they participated in a seemingly endless series of publicity appearances, including television broadcasts with such celebrities as Oprah Winfrey and Matt Lauer and news coverage in *Marie Claire*, the *New York Times*, and innumerable other publications across two continents. A whole set of online blogs devoted to the case sprang up, all but a few devoted to discussions of Amanda's innocence. Thousands of people rallied behind her parents' efforts, and a Washington state senator went so far as to "convey her concerns to Secretary of State Hillary Clinton."

The Knox/Mellas family based their argument for innocence on three pillars. First, Rudy Guede was fingered as the lone and single murderer. The campaign fought long and hard to prove that the break-in may not have been staged but was perfectly real, that someone could have climbed in through Filomena's window even though it was jagged with smashed glass and high off the ground, and that a single attacker could have caused all the injuries inflicted on poor Meredith. Second, Amanda's various missteps, her false accusation of Patrick Lumumba, and her shifting stories—she was not there, she was there, she was not there—were explained away as consequences of police pressure, coercion, and mistreatment during her interrogation.

Third, and most significantly, the scientific evidence against Amanda was discredited. Mixed drops of Meredith's blood and Amanda's DNA found around the cottage were said to be normal, as Amanda lived there and her DNA would have been all over the place, so Meredith's blood may have simply fallen upon it. This argument was used even for the mixed trace found on the floor of Filomena's ransacked room, even though it would have to mean that either Filomena's whole floor was covered with Amanda's DNA, or that Amanda had merely left a small trace here and there, but was remarkably unlucky when the murderer chose one of those precise spots to drop a bit of blood on. Another possibility raised was that Amanda's DNA might have been tracked into the room later and deposited on the bloodstain by the white-suited police investigating the crime scene. All of these arguments and more were brought forward to provide possible explanations for the evidence.

But there remained the damning knife, the most dangerous piece of evidence against Amanda, and one that legitimately raised doubts in the minds of many who were following the case. If Meredith Kercher's DNA was found on the blade of a large and murderous kitchen knife taken from the home of Amanda's boyfriend, a place where Meredith had never set foot, then one may reasonably ask how it got there. Or, as the innocence campaign insisted, whether it was ever really there at all.

ON FEBRUARY 3, 2009, Amanda's aunts Christina Hagge and Janet Huff appeared on *CNN Headline News,* where they were interviewed by star CNN anchor Jane Velez-Mitchell, famous for her irreverent comments on high-profile cases. From the transcript of that show, it is easy to perceive that Velez-Mitchell, while welcoming, is also challenging in her attitude and does not hesitate to ask pointed questions.

VELEZ-MITCHELL: If you're saying that Amanda wasn't involved, who is involved? This other man who was a native of the Ivory Coast? What happened?

HAGGE: Rudy Guede has been charged and sentenced to 30 years for participating in a crime. Amanda and Raffaele are completely innocent. They were home that evening together enjoying a very quiet evening, and they had nothing to do with this.

VELEZ-MITCHELL: But what about the knife? What about the knife?

HUFF: What about the knife? The knife has already been thrown out because A, it does not fit the size of the wounds that were made on Meredith and, B, yes Amanda's DNA is on the handle, it's a cooking knife that she's used at Raffaele's house, but the DNA that's on the tip of that blade is a *less than 1% match to Meredith. It could be more yours or mine than hers* [emphasis added].

HAGGE: And the DNA is not on the tip of the blade. It is not blood DNA. It is—the DNA is on the back side of the blade.

VELEZ-MITCHELL: So what you're saying is that she used a cooking knife and it was in the home of her then boyfriend. And the police went to that house and took the cooking knife and said that was the murder weapon, when it doesn't match up and it wasn't at the scene of the murder.

HUFF: That is what the prosecutor is alleging, yes.

"What a stupid prosecutor," one might well think upon listening to this exchange. On the strength of this feeble scientific result—a trace of DNA on the blade that has less than a 1% chance of belonging to the victim—he actually has an innocent girl and her boyfriend arrested when the real culprit is already in jail? What a terrible judicial error!

Edda Mellas also appeared on *The View,* a show hosted by famed actress Whoopi Goldberg, legendary television anchor Barbara Walters, and a handful of other stars, and devoted to receiving the celebrities of the day. Edda's visit immediately followed an interview with charming Indian actor Dev Patel, star of the hit film *Slumdog Millionaire.*

Sitting at the round table where the hosts receive their guests for animated discussion, Edda fielded questions about every aspect of the case, including the knife. Yes, it was found at Amanda's boyfriend's house, she explained. "There is a very poor chance that the DNA found on it actually belongs to the victim," Edda stated.

On NBC's *Today Show* with Matt Lauer, Amanda's parents brought her younger sister Deanna along with them. Relaxing on the sofa in front of a coffee table decked with pink flowers, while Deanna expressed such sentiments as "They don't really like her there because she's a pretty girl" and Edda added that "Amanda just wants this to be over so she can be home," it was Curt who undertook to point Matt Lauer's attention to an image of the huge knife, explaining that "while Amanda's DNA is prominent on the handle, *the level of Meredith's DNA found on the blade falls below even the level for a poor match* [emphasis added]."

Millions of viewers were being told, repeatedly, that there was a 99% chance that the cells found on the knife blade were not Meredith's.

IN ITALY, accused criminals may choose to undergo a fast-track trial rather than the detailed, full-length version. In return for saving the state the costs of a full trial, they may be given a sentence reduction. Rudy Guede chose the fast-track trial, and on October 28, 2008, he was convicted and sentenced to thirty years in jail for the murder of Meredith Kercher.* That same day, Amanda Knox and Raffaele Sollecito were ordered to stand trial together; they chose the standard-length trial, which in Italy can last a year or more. Their trial began on January 16, 2009.

---

*Later reduced to sixteen years on appeal.

As might be expected, the knife played a key role during the trial, and the cool, dark-haired Dr. Stefanoni underwent a grueling session of questioning and cross-questioning that lasted for two full days. One of the first questions Stefanoni faced was whether the DNA that produced the famous result so similar to Meredith's might not have come from the knife at all, but from some molecule of Meredith's samples left in the machine from a previous test or floating in the air of the laboratory—in other words, whether it might not have been a result of contamination. But Stefanoni's testimony in court on this point was absolutely firm. The judges' report states that Stefanoni

> excluded the possibility that in the machine used for the analyses of the various traces, any secondary deposits might form from which it would be possible to transfer DNA onto other traces. With respect to this, she stated that the machine is equipped with a security system that prevents such an occurrence. With respect to laboratory contamination, she stated that . . . she was not in possession of any data referring to such contamination, and emphasized that if all the procedures associated with good laboratory procedure are applied, the possibility of such contamination is excluded.

The second and more difficult question was whether the avant-garde methods Stefanoni used, such as modifying the machine's sensitivity to test an LCN sample without dividing it into more than one part, should be permitted as evidence at all in a court of law, given that they have not yet been subject to rigorous international scientific testing. Here are the words with which Dr. Stefanoni defended the reliability of her results.

> If an analysis is performed following all the parameters of reliability and proper laboratory procedure, with the due positive and negative controls and the due precautions of wearing the single-use gloves and everything else which is indicated in proper laboratory procedure, then I can be tranquilly certain of getting a result, even with a very tiny quantity of DNA. Therefore I can use the DNA for a single analysis even without being able to repeat that analysis, even if I wanted to. And that analysis is absolutely valid; it has no reason to be put in doubt, as long as the data is absolutely readable and interpretable.

The final question raised by the defense was whether it was possible that the electropherograms showing Meredith's DNA and the one showing the DNA from the knife did not actually coincide. In her testimony, the experienced geneticist simply responded that "comparing the two electropherograms, I saw nothing different, and nothing additional, which could have led me to think that the profile did not belong to the victim but to some other person, known or unknown, but another individual." What Stefanoni saw is the same thing that anyone can see by looking at the graph obtained by superimposing the DNA electropherogram of the knife DNA on top of Meredith's DNA: they are identical.

The court accepted Stefanoni's claim that the DNA on the knife belonged to Meredith and concluded on this basis that the knife found in Raffaele's house, with Amanda's DNA on the handle and Meredith's DNA on the blade, was the murder weapon responsible for inflicting the massive lethal wound to Meredith's neck. Together with the scattered pieces of evidence that fit together to make a bigger picture, they concluded that Amanda and Raffaele had been present at the cottage on that fateful night, and had joined Rudy Guede in an aggressive attack on Meredith that turned fatal.

On December 4, 2009, almost a year after the start of their trial, Raffaele Sollecito and Amanda Knox were convicted of murder. Raffaele was sentenced to twenty-five years in jail, and Amanda, because she had accused an innocent man of murder, to twenty-six.

AMANDA'S AND Raffaele's lawyers immediately lodged an appeal. There were some grounds: by some unfortunate coincidence it seemed as though each and every piece of evidence used to convict the pair was flawed in some way or another. A bloody footprint on the bathmat seemed to fit Raffaele's foot, but the identification could not be certain. His DNA was found on the ripped-off clasp of Meredith's bra, but that clasp had been forgotten in the murder room during the first forensic inspection. Between then and the time it was collected from the floor forty-six days later, objects had been carried in and out of the room, bringing a possibility of contamination from the forensic technicians' shoe covers every time. A witness who claimed he saw the pair talking animatedly in a little square overlooking the cottage gate the evening of the murder said it was the night of Halloween instead of November 1. Another

claimed, a year after the murder, that he had seen Amanda in his little grocery store at eight o'clock in the morning on November 2, but it turned out
that he had failed to identify her when the police showed him photos of her
in the days following the murder. Two witnesses, in addition to Rudy Guede,
told of hearing a terrible scream from the cottage on the night of the murder,
but none of them could really say what time they had heard it.* A broken-
down car had been parked in front of the gate leading to the cottage from
10:30 p.m. to about 11:30 p.m., during which time a tow truck had come to
make repairs, yet none of the people involved had seen a light or heard any
sound from the cottage or spotted anyone going in or out.

In the end, some of the most damning evidence against the couple came
from their own fabrications and contradictions: Amanda's accusation of the
innocent Lumumba; her claim, later retracted, that she had been in the
house when Meredith was murdered; Raffaele's statement, also later retracted, that Amanda had left his house and gone out alone that evening.
Then there was their claim that they had slept straight through the night
until 10:00 a.m., when in fact Raffaele's father had called him at 9:30 in the
morning, and worse, Raffaele's computer showed continuous use setting up
a playlist and listening to music from 5:30 to 6:10 a.m. Also, a text message
from his father late in the evening showed up on his phone after 6:00 the
next morning, indicating that he had turned his phone on at that time as
well.†

But Amanda's and Raffaele's lawyers repeated untiringly that none of
these discrepancies provided any proof of murder, and they left no stone un-
turned in their effort to discredit each piece of evidence one by one. It began
to seem as though the outcome of the appeal trial hung on a single thread:
the DNA, both Meredith's on the knife and Raffaele's on the bra clasp.
Drowning in the conflicting and strongly expressed opinions of the expert

---

*In a Skype call with a friend before his arrest, Rudy Guede mentions this terrible scream
and his certainty that people outside the house must have heard it, and he places it at 9:20
or 9:30 p.m. But since he also claims that Meredith returned home at 8:20 or 8:30 p.m., his
timing cannot be considered reliable.

†In his book, Raffaele explains that he "had been up several times in the night—listening to
music, answering e-mail, making love," but no e-mail message he wrote to anyone that night
has been found.

witnesses who had been hired by both sides in the first trial, the appeal judge, Claudio Pratillo Hellmann, decided to call on a team of independent experts, university forensic geneticists unconnected to the case, to make a final judgment on the quality and reliability of Stefanoni's work. They chose Professor Carla Vecchiotti and Dr. Stefano Conti, both of Rome's highly reputed La Sapienza University. Conti and Vecchiotti studied the knife and bra clasp in their lab; they also examined all of the records from Stefanoni's lab pertaining to the analysis of the two items. They submitted a report on their analysis to the court at the end of June 2011.

The day the experts' report came out was the day that the trial completely changed direction. The experts lambasted the on-site forensic work. They showed a film of the white-garbed technicians lifting samples of blood and hair from the crime scene, focusing on smudges of dirt on the tips of the single-use latex gloves and on the handing of evidence back and forth before bagging it. As far as Raffaele's DNA on the bra clasp was concerned, they deemed it unreliable as evidence because of the possibility that his DNA could have been tracked in by the dirty feet seen on the video going in and out of the room, despite protocol saying that shoe covers had to be changed every time. Although none of Raffaele's DNA was found anywhere in the house except on a single cigarette stub in an ashtray, making it difficult to see where it could have been tracked in from, the two experts' searing criticism of inspection techniques raised some doubt about the validity of the evidence. Furthermore, they pointed to the electropherogram of the DNA from the bra clasp, which, while it contained a clear profile of Raffaele, also contained a rather large number of peaks that seemed too significant to be mere background noise, indicating, rather, some contamination by further unidentified contributors to the sample. Despite the commonly accepted guidelines for interpreting such peaks, Stefanoni had chosen to consider these extra peaks as background noise, stating that she saw only Meredith's and Raffaele's genetic profiles on the bra clasp. Her conclusion was now rejected outright by the experts, who did not pause to explain why, apart from Meredith's, a more significant quantity of Raffaele's DNA was present than anyone else's. They deemed the bra clasp evidence unreliable.

The knife posed a subtler problem. To start with, Conti and Vecchiotti attempted to make a new test, but they found that the quantity of material remaining on the knife was so small as to consist of only a couple of individual

cells, if any. At this point they could have done some tests to quantify it precisely, but they declined to do this on the grounds that even if some cells
were there the sample would be even smaller than the one Stefanoni had
worked with. If LCN results were deemed unreliable by the scientific community and thus unfit for use in court, then any new results would be so as
well, in which case testing the new sample would be useless.

Although they were just as scathing about the collection procedures used
during the forensic inspection of Raffaele's house, including the manner in
which the knife was found, collected, bagged, and transferred to Stefanoni's
lab, the experts were unable to explain how, of all the possible traces of DNA
that might have contaminated the knife during this process—DNA from
the handlers, from other objects in Raffaele's house, or from lab technicians—of all people, it was the DNA of Meredith, who had never been near
either Raffaele's house or the lab, that landed there. It was all very well
claiming contamination, but nothing from the collection procedure—from
kitchen drawer, to lab, to swabbing, to DNA testing machine—could be
blamed for the presence of *Meredith's* DNA.

The experts, however, did raise one point in the procedure where it
seemed as though there was a real chance of contamination. When a very
small sample is run through a machine where good-quality samples from
the same person have been run previously, it is possible for the good-quality
sample to "rub off" on the tiny one, thus contaminating it. Stefanoni had
testified that the knife sample had been run "somewhere in the middle of a
series of 50 or 60 samples of Meredith's DNA." During their court testimony
on July 25, 2011, Conti and Vecchiotti stressed this point. Stefanoni, they
said, should have carried her LCN sample to another lab for testing so as to
avoid this problem. Not having done so left her results open to question.

Five days later, on July 30, the prosecution had their chance to interrogate the experts in court. They had brought in a record of all the tests done
in the laboratory in the days preceding the knife DNA testing.

"Do you know when the last sample of Meredith's DNA was tested in
the lab, prior to the knife DNA testing?" the prosecutor asked the experts.
No, they responded. They had studied only the reports of the knife testing,
not testing of other samples. They didn't know. It turns out that it had been
six days earlier. "The machine was used and flushed out repeatedly in between the previous sample from Meredith and the knife sample. Is it still

possible that a sample from her DNA contaminated the one from the knife?" "No," the experts admitted. A six-day delay was too long for that type of contamination to have occurred.

As a last resort, Conti and Vecchiotti held fast to their statement that the sample was too small for the results of the tests to be considered reliable. To support their position, they cited a number of scientific publications warning against the use of LCN DNA samples in court for the purposes of conviction. They did not make any assessment of the reliability of the result in the particular case of the knife electropherogram despite the low level of background noise that left the genetic profile clearly visible.

The prosecution submitted a formal request for the new, tiny sample swabbed from the knife by Conti and Vecchiotti to be submitted for analysis. In court on September 5 and supported by expert prosecution witness Giuseppe Novelli, Stefanoni explained that newer generations of DNA analysis kits existed in 2011 that had not been available in 2007, and these new kits could give results on samples as small as a couple of cells. She wanted a new analysis performed to confirm that her previous work was correct. The prosecution agreed and asked the judge to order the new tests.

The request was rejected by Judge Hellmann on September 7. On the day that this decision was announced, even Barbie Nadeau, the author of *Angel Face: The True Story of a Student Killer,* and one of the most unambiguously "guilter" journalists following the case, tweeted directly from the courtroom: "Looks like Knox will walk." Indeed, it was difficult to interpret the judge's decision any other way.

Nadeau was not wrong. On October 3 the jury deliberated for ten hours, but seemed pleased and full of smiles when they finally returned, and the judge read out the decision that overturned the verdict of the first trial, declaring that Amanda and Raffaele had not committed the crime of which they were accused and setting them free.* Their verdict was in flat contradiction with the Italian supreme court verdict decreeing that Rudy Guede had not acted alone, and with autopsy reports that clearly indicated that Meredith had been attacked by more than one person. But above all, the verdict was in contradiction with the DNA found on the knife.

---

*Amanda was essentially sentenced to time served for her unwarranted accusation of Patrick Lumumba.

How did the appeal court justify rejecting the knife DNA and refusing the prosecution's request for a retest using new technology? In some countries we would never know the answer to that question, but in Italy judges are required to produce a document called the "motivations" for the sentencing, which explains the reasons for their decisions in detail. Here is the passage in which Judge Hellmann justified his choice concerning the knife:

> We deduce that for our purposes, the result obtained by the Scientific Police cannot be accepted as reliable, since it is the product of a procedure which did not follow the techniques indicated by the International Scientific Community, or in any case its reliability must be seriously weakened, so much so as to make it necessary to find confirmation in other elements independent of the scientific analysis.
>
> This also explains why the expert team did not proceed farther in analyzing the sample that it collected from the blade of the knife: the quantity was found to be again LCN, and altogether insufficient to make two separate tests* possible, so that if they had proceeded further, the court-appointed experts would have committed the same error as the Scientific Police. And on the other hand, it seems clear from the ideas explained above that because the necessity of dividing the sample into two or more parts holds for every single trace, its aim being to guarantee the reliability of the result of the analysis of that trace, it is not by analyzing two different traces that are both LCN, without treating either of them with the proper procedure to guarantee the result, that one can think to make up for the lack of repetition in the procedure for each single trace: the sum of the two results, both unreliable due to not having been obtained by a correct scientific procedure, cannot give a reliable result.

Here, Hellmann is making a statement about experiments whose outcomes are reliable with a particular percentage of certainty. Let us say, for example, that we run an experiment whose result has an X% chance of being

---

*By this phrase the judge is referring to the standard process of performing two separate tests on the new sample by dividing it into two parts.

correct. We then run the experiment independently a second time and obtain roughly the same result, again with an X% chance of being correct. What Hellmann is saying is that the fact of having run the experiment independently two times and obtained the same result twice does *not* increase the reliability of the result. "The sum of the two results, both unreliable . . . cannot give a reliable result." This sentence shows a complete misunderstanding of the probabilistic result of considering two separate results from two performances of the same test. It is precisely the situation explained in the introduction to this chapter with the weighted coin.

To show how wrong Hellmann's reasoning is, here is another example, this time with an initial probability of correctness of approximately 80 or 90%. In fact, it is difficult to give any precise assessment of the probability that the knife electropherogram represents Meredith's DNA. Visually, it seems quite certain that it does. But the advantage of an estimate of 80–90% is that such a probability is relatively convincing, but certainly not beyond a reasonable doubt. Let us now consider what can be deduced, mathematically, if you do two tests and obtain similar results.

The new example is similar to the one in the introduction: you are given a coin that is either fair or biased to fall on heads 70% of the time, and after a certain number of tosses you must decide whether or not the coin is biased. Now, though, let us suppose that a test consists not of a single coin toss, but of ten.

You do a first test and obtain 9 heads and 1 tail. Knowing that your coin is either fair or biased, the probability that the coin is fair given this outcome is about 8%, or that it is biased, about 92%. Pretty convincing, but not enough to convict your coin of being biased beyond a reasonable doubt.

You do a second test, and this time you throw 8 heads and 2 tails. Now the probability for a fair coin is about 16%, for a biased coin about 84%. So the naïve thought might be that you haven't gained any certainty from this second test; if anything, the 92% certainty that you had before has now been diminished a little. This is obviously the line of reasoning going on in Hellmann's mind when he wrote the passage above.

But if you think about it differently, what you've really done is throw the coin 20 times and get 17 heads and 3 tails. Using the exact same probability calculation as in each separate case—the one explained in the introduction—now yields a probability of 98.5% that the coin is biased! There is no legal set numerical threshold for reasonable doubt, of course, but 98.5% is a lot closer to that elusive notion than 92% or 84%. Thus, running a test that is only moderately reliable twice and getting the same result may indeed, in total, yield a very reliable result.

THE APPEAL trial verdict is presently being appealed to the Italian supreme court, which has the right to either confirm the appeal verdict or cancel it and order a new appeal, during which the knife DNA might or might not be retested. What will happen, only the future can tell.

In the meantime Raffaele and Amanda have returned home, one to Bari, Italy, the other to Seattle, and both have undertaken to write books chronicling their experiences and proclaiming their innocence. Whoever killed poor Meredith will probably never tell the complete story. What we do know is that by using flawed scientific reasoning to reject a technically possible retest of the knife DNA, Judge Hellmann missed a major opportunity to get at the truth.

MATH ERROR NUMBER 5 »
THE BIRTHDAY PROBLEM

THE *BIRTHDAY PROBLEM* is a classical probability puzzle that asks the following question: how many people do you have to put in a room for there to be a 50-50 chance that two of them share the same birthday?

Before reading ahead, try to guess the answer. Most people will correctly intuit that if 2 randomly selected people are in a room, the chance that they happen to share the same birthday is 1 in 365 (discounting February 29 births!). But if 3 people are in a room, the chances of any shared birthdays increase to about 3 in 365. With 4 people in the room, the chances move up to about 6 in 365, and with 10 people, they increase to a surprising 52 in 365, or over 14%. At 23 people in a room, the chance that at least two of them share a birthday is very close to 50%.*

This final result seems very counterintuitive to most people, who tend to guess that to get a 50% chance of a shared birthday, you would need about 183 people, or roughly half of 365. At just 23 people in a room, most people

---

*To calculate this, it is easiest to compute the opposite probability: that everyone in the room has a different birthday. For 2 people in the room, it is 364/365, as person 2 has 364 "free days" for his birthday (the ones that are not person 1's birthday). For 3 people, this number has to be multiplied by 363/365, as person 3 now has 363 free days, and for 4 people, the number has to be multiplied by 362/365. Continuing this calculation on an ordinary calculator leads to the probability of just slightly under ½ for 23 people all having different birthdays, meaning just slightly over ½ for two or more of them to share.

tend to think it is far more likely that they will all have different birthdays—yet this is not the case.

The funny thing is, people often offer the same answer for a quite different question, which has to do with specifying a particular date: how many people do you need to put in a room for there to be a 50-50 chance that one of them has a birthday on January 1?

The answer to this question is not 183 either, but 253. The reason is that as you keep adding people to the room, the probability that two or more of them share birthdays increases. Given that 364/365 is the chance of one person having a birthday not on January 1, $(364/365)^n$ is the chance of n people not having that birthday. Since we want that chance to be less than 50%, we look for the smallest n such that $(364/365)^n$ is less than 1/2, which is 253.

The unexpectedly large difference (23 people versus 253 people) between the answers to two similar-sounding questions is a trap that one can fall into quite easily, as the following case illustrates.

The Case of Diana Sylvester: Cold Hit Analysis

In the early morning of December 22, 1972,[*] twenty-one-year-old nurse Diana Sylvester walked the few blocks home from her night shift at the San Francisco Medical Center of the University of California, reaching her Sunset District apartment at around 8:00 a.m. Her roommate, Patricia Walsh, also a nurse at UCSF, worked day shifts starting at 7:00 a.m.; when Diana got home that morning, Patricia had already left. What happened next has never been fully elucidated.

Shortly after 8:00 a.m., Diana's landlady, Helen Nigodoff, who lived in the apartment below, heard loud noises coming from upstairs. The thumps and screams went on for a good twenty minutes before she finally decided

---

[*]The date of Leila's birthday—viz. the birthday coincidence problem.

that she had better see what was going on. It wasn't the first time there had been disturbances in Diana and Patricia's apartment; they had many friends, male and female, and Helen had gone up to complain about the noise three weeks earlier. But on this morning something didn't sound right.

Helen rang the downstairs doorbell to Diana's apartment, but since the door was already open, she didn't wait but hurried up the flight of stairs toward the apartment's main entrance. A man standing in the doorway startled her. Helen asked what was going on. "Go away, we're making love," he snarled aggressively, and

Diana Sylvester

she quickly turned back down the stairs. From the safety of her own place, she called the police and told them that something violent and frightening was going on upstairs. She managed to get a good look at the stranger as he ran down the stairs and out of the building.

Minutes later the police arrived and rushed up the stairs to Diana's apartment, where the door was still wide open with no signs of having been forced. There they were met with a tragic sight. Under a brightly lit Christmas tree with gaily wrapped gifts heaped underneath it, Officer John Forbes and his colleague Inspector Kenneth Manley found Diana's naked body. Her clothes were in a pile next to her, and she had two bleeding wounds in her chest. As the postmortem revealed, the killer had forced her to perform oral sex on him, strangled her, and then stabbed her twice in the heart.

FINGERPRINTS IN the room and sperm samples taken from the victim's body were stored by the police, but DNA analysis did not yet exist. The only clue to the identity of Diana's killer was a statement given by Helen Nigodoff, who described the man she had seen as "white, medium height, heavy-set, chubby, curly brown hair, beard, mustache, with a clean-cut appearance."

As in the Janet Collins case (see chapter 2), the police were reduced to searching the neighborhood for individuals who fit such a description. They managed to find one quickly. His name was Robert Baker, and he was a thirty-two-year-old street artist who lived in his Volkswagen van. He had escaped from a mental institution a month earlier. He was the primary suspect in a rape that had occurred two weeks before Diana's murder, merely four blocks away from her house; that victim had actually identified him. She hadn't been killed, but Baker had threatened her, saying, "I can rape you now or after you're dead." Also, as in Diana's case, the front door to the victim's flat hadn't been forced; he had persuaded her to open the door to him on some pretext. Police records also show that four days after Diana's murder Baker had harassed a young girl and her nanny, following them to their home a few doors down the street from Diana.

In their search for a possible link between Robert Baker and Diana Sylvester, the police hit on another clue. A week after the murder, Charlene Nolan, another nurse at UCSF and friend of Diana's, told the police that on her way home on the morning she died, Diana had planned to stop and buy a candle from a street artist at Millberry Union Plaza. Patricia, Diana's roommate, confirmed that there was indeed a new candle in the flat. Charlene knew the street artist; his description did not correspond to the man seen by Helen Nigodoff, but on the other hand, the police discovered that Robert Baker had been selling drawings not far from where Diana bought her candle, and could easily have seen her and followed her home.

When they searched Baker's van, police found mail stolen from Sunset District mailboxes and a parking ticket with drops of blood on it. The blood type was O, the same as Diana's, but because it is also the most common blood type, the finding could not be considered conclusive. Given the limits of forensic science at the time, no further information could be extracted from the ticket. Unfortunately, it was later mislaid or destroyed.

On January 11, the police organized an identification lineup that included Robert Baker. Although records do not state precisely which witnesses were asked to take part in the identification process, it can be inferred that Helen Nigodoff was one of them. She must have failed to pick Robert Baker out of the lineup, though, because the case came to a close for lack of evidence, and Baker was never charged for Diana's murder. He died in 1978, and Diana's case files sat, among all the other unsolved "cold case" files, gathering dust for more than thirty years.

IN 2003 the San Francisco Police Department received a grant to use new DNA technologies on a set of old, unsolved cases in which usable DNA evidence had been successfully preserved. The new method consisted of taking a DNA sample from a piece of evidence shut away in the cold case files and running it through a large database containing the DNA of thousands of known California criminals in order to search for a match.

Although it was over thirty years old, Diana's case file contained a slide with a swab of sperm taken from Diana's dead body. For this reason it was chosen as one of the cold cases to be reexamined. Unfortunately the sperm sample was extremely degraded, so that only a part of the DNA could be read; out of the thirteen genetic pairs (called genetic loci; see the explanation of DNA analysis in chapter 4) that form a complete DNA identification profile, only five pairs and partial indications of two or three more were visible on the electropherogram from the sample.

In DNA analysis, if two graphs show peaks located in distinctly different places, this absolutely precludes a match. But what often happens in the case of degraded DNA samples is that a graph showing only a few clear peaks is matched to a complete sample, and no differences are visible. In other words, every peak present in the degraded sample exactly matches a peak in the complete sample, but it is impossible to know if the missing peaks from the degraded sample would correspond or not. Still, the investigators took the partial DNA profile of Diana's attacker and ran it through the system. They came up with exactly one candidate for a possible match: one person in the whole criminal database whose DNA graph showed peaks in the exact same positions as those of Diana's murderer.

The match was John Puckett, a seventy-two-year-old man from the Bay Area. The reason his DNA profile was in the database was because some quarter of a century earlier he had been convicted on three counts of rape. These all occurred in 1977, and his method of operation had been quite similar in all three cases: he would approach women, pretending to be a police officer, threaten them with an ice pick or a knife, and force them to drive to an isolated area in Marin. There, as two of the women testified at his trial, he raped them; the third was compelled to perform oral sex. Puckett was convicted and sent to prison until 1985. After his release, apart from a misdemeanor battery charge in 1988, for which he was charged but not convicted, his record stayed clean. In 2003 he was living with his wife in a mobile home and was now an elderly, ill man in a wheelchair.

On October 12, 2003, San Francisco Police Department homicide inspectors Joseph Toomey and Holly Pera knocked on the door of Puckett's home. He answered the door, clutching a urine bag in his hand—he had difficulty walking, having recently undergone triple-bypass surgery. The inspectors interrogated Puckett for over an hour. Puckett denied having ever met Diana, having had sex with her, or having been inside her house. He offered to provide Toomey with a fresh DNA sample; the offer was accepted and confirmed the match with the cold case sperm sample. At the time of the murder thirty-three years earlier, no one had connected John Puckett with the crime, and he had never been a suspect.

On the evidence of the DNA match, together with the corroborating facts that he had lived in the area at the time of Diana Sylvester's killing and that he was old enough to have been able to commit it in 1972—but with no direct evidence of his involvement—Puckett was arrested and charged with murder. According to the officers who arrested him, he behaved "like a gentleman" when they took him away. He turned to his wife and said, simply, "I guess I won't be seeing you anymore."

Toomey interviewed Puckett again and searched his home for old pictures. One picture dated back to Christmas 1972, three days after Diana's murder. The picture showed that at that time, Puckett was curly-headed, had a great deal of facial hair, and was significantly overweight. The similarity to Helen Nigodoff's description ("white, medium height, heavy-set, chubby, curly brown hair, beard, mustache") was plain. But Puckett continued to deny knowing anything about the crime.

AS WE saw in the Amanda Knox case (see chapter 4), difficulties with identifying DNA can arise when the samples are too small, when they are mixed samples from more than one person, or when they are degraded. When a degraded sample shows only a few clear peaks rather than the full set of 13 genetic pairs (loci) ordinarily used by forensic biologists, even a good match of the peaks that are present with the DNA of a given individual cannot be considered complete identification. Indeed, it is common for different members of the same family—and even perfect strangers—to share a few peaks.

The California state law usually requires a minimum of 7 loci to even consider running a degraded sample through its database of offenders.

By studying large DNA databases, the FBI has calculated something called the random match probability (RMP), which measures the likelihood that two unrelated people have any given number of matching loci. For example, the probability of two strangers sharing 13 identical loci is given as about 1 in 400 trillion. Since the population of the world is only about 6 billion, it is considered a certainty that if two samples match at 13 loci, they must belong to the same person.

Each gene locus of the 13 pairs usually considered has a fixed probability called the random match probability—which has been measured precisely—of occurring in a specific position for any given individual. The RMP values differ somewhat from gene to gene, but the average is roughly about 1 in 13. In other words, any given configuration for 1 of the 13 genetic loci used for identification will be held by roughly 1 out of 13 people, or about 7.5% of the population.

Given the fact that the 13 gene loci are known to be statistically independent from one another, it is correct to multiply these probabilities together to estimate the proportion of people who will share several identical gene loci. For instance, the proportion of people having 2 given gene loci will be about $(0.075)^2$, meaning about 1 in 177 people; the proportion of people having 3 specified loci is about $(0.075)^3$, meaning 1 in about 2,370 people, and so on. For greater numbers of specified loci, the proportions become very small: 1 in 177 billion for 10, 1 in 2 trillion for 11, and 1 in 31 trillion for 12.

The RMP values for each gene pair calculated by the FBI are more sophisticated than this figure of 7.5% and therefore generate more precise probabilities, but the overall numbers are actually quite close to the estimates given here. For 9 loci, for example, the RMP is given as 1 in about 13 billion. This means that given a specific set of 9 pairs, we can expect to find about 1 person in 13 billion who will have those 9 pairs in exactly the specified locations. This published figure of $1/13,000,000,000$ is very close to our estimation of $(0.075)^9$.

These numbers are the product of well-documented science and were deemed indisputable—until one person began to question their correctness.

When the task of participating in Puckett's defense fell to Bicka Barlow, an attorney with a background in forensic genetics who worked for the San Francisco Public Defender's Office, she sat up and took notice. Bicka had been worried for quite some time about how DNA evidence and the FBI's RMP were being used to search databases for cold hits. She believed that this method was leading to the arrest of innocent people due to miscalculation of probabilities. When Puckett's case landed on her desk, she saw an opportunity to investigate the problem in depth.

Bicka had been particularly struck by the research of an Arizona state employee named Kathryn Troyer who had carried out a major statistical study in 2001. Troyer ran a series of tests on a DNA database containing over 10,000 profiles. Given the 1 in 13 billion statistic, one might not expect to see any matches at 9 places in such a small sample. Yet Troyer did find one such pair of individuals in the database: two unrelated individuals who shared 9 identical genetic loci.

One example may not prove much, but as the years passed and the database increased, Troyer kept performing her tests again and again. In 2005, using over 65,000 profiles, she found 122 pairs with 9 matching loci and 20 pairs with 10 matching loci. This appeared to show that in spite of minuscule probabilities of such matches occurring in the general population, they actually do occur and not that infrequently.

In preparing John Puckett's defense, Bicka pounced on the results of Troyer's study. In her view these results could only mean one thing: the FBI's RMP statistics must be wrong. If as many as 122 pairs could match at 9 loci in a sample of just 65,000 people, then what could that 1 in 13 billion statistic mean? "[So many] matching pairs at 9 loci is an incredible fact. . . . The State has information that they're not providing to the defense that says that, in fact, their statistical analysis is wrong and it could be wrong by orders of magnitude. . . . I could have gotten a statistician to calculate a probability . . . it's almost—it's improbable."

What Bicka Barlow was expressing was an intuition that while the RMP figures of 1 in 13 billion give the impression that 9-locus matches are incredibly rare, database experiments show that actually they are fairly common, and these two facts appear to contradict each other. In the appellant's brief (the legal document preparing Puckett's appeal), this seeming contradiction is noted more bluntly as one of the major grounds for appeal: "The trial court erroneously prevented appellant from challenging the prosecution's statistical analysis by presenting evidence of matching pair studies to demonstrate that the probability of obtaining a false match was much higher than the Random Match Probability would suggest."

Should we have "serious concerns as to the inculpatory force of the RMP statistics" as the appellant's brief suggests? Are they reliable enough to provide evidence against anyone at all? How many innocent people have already been locked away behind bars by those frighteningly tiny probabilities?

THE ANSWER lies in a simple statistical calculation, correctly identified in the respondent's brief (response by the State of California to the appeal) as nothing other than the famous "birthday problem."

As a general matter, the Arizona results are merely an illustration of a well-known but seemingly paradoxical mathematical concept known as the "birthday problem." The birthday problem asks, "What is the minimum number of people in a room for the odds to be better than 50% that at least two will share the same birthday, no matter what day it is?" This question assumes that the rarity of any particular birthday is 1 in 365 (i.e. its random match probability). The correct answer, although belying common sense, is a mere 23 people. In other words, pairs of relatively rare events are *expected* in relatively small databases, as long as one is not looking for a specific target event. The solution is based upon the fact that with only 23 people, 253 separate pairwise comparisons can be performed, making it likely that some birthdays will occur twice. A key point is that the question does not ask how many people would have to be in a room to expect to find a *particular* birthday represented (e.g. January 1). The rarity of any given, predetermined, event (e.g. a January 1 birthday, or a target DNA profile in the DNA database context) remains unchanged.

The respondent's brief quotes Bicka's remarks on the Arizona data during the trial: "If you are going to find two people that match at 9 loci in a database of 10,000, that says a lot about whether or not somebody else out there shares Mr. Puckett's profile, and it flies in the face of the common sense understanding of what that random match probability means!" They correctly identify this remark as a typical birthday problem fallacy, confusing the possibility of a match between *any two people* in the database, which is reasonably large, and the possibility of a match with *one particular* DNA sample given beforehand, which is minuscule!

In their words: "Were the issue framed instead in terms of the birthday problem, defense counsel's argument would seem to be that observing one or more pairs of birthdays when comparing everyone in a relatively small room says a lot about whether or not somebody else out there shares, for example, a March 5 target birthday.* This, of course, is false."

The application of this reasoning to the problem of whether or not the Arizona database results contradict the FBI's random match probability is not calculated explicitly, although the respondent's brief expresses the right idea. The question we need to answer is the following: Given the FBI's random match probability statistics, how many 9- (or 10-, 11-, 12-) locus matches would we expect to find, on average, in a database of 65,000 profiles?

What people forget when they see the "1 in 13 billion" figure is that what is being measured is not a number of individuals, but a number of pairs of matching individuals—and that in any population, the number of pairs is far higher than the number of individuals.

In a population of N people, there are $N \times (N - 1) / 2$ pairs of people. When searching for a 9-locus match, every individual is tested against every other. Furthermore, there are 715 different ways to choose 9 loci out of the 13 usually considered by forensic geneticists, so that "a match at 9 loci" can mean a match at any one of these 715 choices of 9 loci.

In table 5.1, we consider the databases of sizes 10,000, 60,000, and 65,000 that Troyer studied. The number of tests for matches performed in each case is given by $715 \times N \times (N - 1) / 2$, where N is the size of the database; indeed, each of the $N \times (N - 1) / 2$ pairs of individuals is tested for a match at each of the 715 different choices of 9 loci among the 13. The number of matches predicted by the RMP is obtained by multiplying this total by the FBI's predicted frequency of 1 in 13 billion. The final column shows the number of 9-locus pair matches that Troyer actually found.

---

*This date, explicitly given as an example in the brief, just happens to be Coralie's birthday; like the coincidence of the date of the murder being Leila's birthday, this illustrates the fact that birthday matches—like DNA matches—occur more frequently than one might intuitively suppose.

TABLE 5.1

Database Size	Pairs Tested	Expected 9-Locus Pair Matches	Matches Found
10,000	35,746,425,000	2	1
60,000	1,286,978,550,000	98	90
65,000	1,510,414,262,500	116	122

From this table it appears clearly that the expected answers are close to the actual answers that Troyer found. There is, in fact, no contradiction with the 1 in 13 billion random match probability. There is only the "surprise" of realizing that the 1 in 13 billion figure refers to 1 in 13 billion *pairs*, and that the number of pairs is large for even quite a small population; indeed, there are already over 35 billion pairs in the smallest population of 10,000 individuals. This is exactly the principle behind the birthday problem: there are 253 pairs of people in a group of 23 people, and thus the chance of finding a common birthday is quite large—as we saw, greater than one in two.

What Bicka Barlow and Puckett's defense team tried to argue is that the Arizona data shows that there are really a great many matches out there, many more than the RMP figures might lead one to believe. What they are ignoring is that these surprising numbers of matches occur when you search among all the pairs in a population for *any possible match of any set of genetic loci whatsoever*. This is entirely different from a cold-hit database trawl, in which one DNA sample is given at the start and you search through the database to find the same one. In this situation the probability of finding a match will be much lower, as is the case with the two different birthday problems explained above.

In conclusion, there is no contradiction at all between the Arizona database findings and the RMP statistics provided by the government. The suspicions raised by the defense here were unjustified and based on an intuition that, although natural and widespread, is faulty.

A SECOND, more complex problem emerged during the preparation of Puckett's trial. The RMP for a match at the five and a half genetic loci visible in the sample from Diana's murderer was calculated as 1 in 1.1 million. The defense challenged this figure by a second argument, this time not so much a challenge to the figure itself but to its meaning. The gist of the argument is this: the figure signifies that if you pick someone off the street, there's 1 chance in 1.1 million that he or she will share the same five and a half genetic loci as Diana's killer. If you search through a large database, this comes down to picking that many people and doing the comparison with each of them, so the chance of finding a "random match"—that is, an individual with the same genetic loci but who is not the source of the original DNA sample—is multiplied by the number of people in the database. In the case of the Puckett cold hit, the database consisted of 338,000 registered sex offenders from California.

A comparable example is the following. Consider a lottery with 1.1 million tickets. If you buy 1 lottery ticket, the probability that the winning number is on your ticket will be 1 in 1.1 million. But if you buy 2 tickets, your chance of winning will double, and if you buy 10 tickets, you're multiplying your chance of success by 10. Thus in this lottery, buying 10 tickets would increase your chances to 10 in 1.1 million, or 1 in 110,000. If you buy 100 tickets, your chance of winning goes up to 1 in 11,000, and if you buy 1,000 tickets, your chance is now 1 in 1,100. What if you go so far as to buy 338,000 tickets? Then your chance of winning would be 338,000 in 1,100,000, which is equal to 338 in 1,100, or just a little under 1 in 3.

Now think of the criminal's DNA sample as the winning number, and a random person from the population as a ticket that has a chance of 1 in 1.1 million of matching. The idea is that if instead of picking 1 random person, you pick 338,000 people, as in the lottery, your chance of finding a match would be just a little under 1 in 3.

Thus while the prosecution seemed to say that there was 1 chance in 1.1 million that Puckett could be someone other than the source of the DNA found on Diana's body, the defense countered that since he was found by a database search of over 338,000 people, there was really about 1 chance in 3 that Puckett could be merely a "random match," and thus innocent.

It's very different to say that some event—Puckett being innocent and merely a random match to the criminal—has 1 chance in 1.1 million of occurring, or 1 chance in 3. In computing the latter probability, Bicka Barlow thought she had the means to make the Puckett trial into a showcase for her theory that cold hits carry a tremendous risk of sending innocent people to jail. In her opinion she had never seen a weaker case against a defendant in a murder trial; apart from the DNA there was no evidence whatsoever against the defendant other than a rough similarity with a thirty-year-old eyewitness description. Bicka was a fighter: her mother was a Holocaust survivor, and Bicka had spent part of her childhood attending the peace rallies that took place around Berkeley, perched on her father's shoulders. Now she thought she had a prime case for victory in the battle against the conviction of innocent people by cold hits. But things didn't work out quite as planned.

BEFORE THE trial even started, the judge decided that it would be too confusing to present the different and seemingly contradictory mathematical arguments in court. Not only did he exclude the use of the 1 in 3 probability figure, but he also ruled that the jury could not be informed that Puckett had been identified as a result of a cold hit from a DNA database. "I'm not inclined to change my opinion regarding what's in and what is out, and so the database search for matches and partial matches, that experimental exercise or line of research, that is out," he pronounced at the preliminary hearing.

Because of the judge's decision, none of this information—which formed the backbone of the case that Bicka was preparing—was ever presented in court. One juror even sent a note to the judge asking just how Puckett had been identified as a suspect. He was told that the information wasn't relevant. "I am instructing you not to guess," the judge wrote in his response.

Instead the jury heard about the evidence collected from the crime scene, about Helen Nigodoff's testimony, and about Puckett's past convictions for assault and rape. They were told how when interrogated about Diana's murder the defendant had responded, "I don't remember this at all," and how he had said the same thing nearly thirty years earlier when faced with the evidence of his other crimes. They learned that he had told the women he attacked that he wanted to "make love"—the same phrase the murderer used when he growled to Helen Nigodoff, "Go away, we're making love." They heard each of Puckett's three earlier victims describe what he had done to her, and how his weapon had left scratches on her neck similar to those found on Diana's dead body. And of course the jury heard about the random match probability of 1 in 1.1 million that Puckett and Diana's murderer could share five and a half genetic loci by chance.

Deliberations began on February 14, 2008. It took nearly twenty hours, spread over several days, for the jury to reach a decision. During this time the jurors contacted the judge with a note: "We know that cold DNA cases were evaluated at some point. Given that the defendant had not yet provided his oral swab, how was he identified as a person of interest?" It seemed clear that the question could not be erased from the jury's minds, and the court finally decided to stipulate that Puckett had been found as the result of a cold hit—only to be foiled by the defense, who protested that such important information should not be revealed to the jury without a complete explanation of its significance. The defense believed that without the proper explanations, the aura of scientific prestige surrounding the term "cold hit" would work against the defendant. Once again, the jury was told that the question was irrelevant and that they should refrain from guessing.

After these lengthy deliberations, they finally reached a verdict and declared John Puckett guilty of first-degree murder. On April 9, he was sentenced to life imprisonment. Declaring his innocence, he appealed the verdict the same day.

NATURALLY, THE main ground of appeal was the exclusion from the trial of any mention of the 1 in 3 chance of the match to Puckett having occurred purely by chance, calculated by Bicka Barlow. From the appellant's brief:

> What the jury did not know, and what the trial court believed the jury did not need to know, was that the real chance of a coincidental match in this case was 1 in 3. This number would have placed the prosecution's case in an entirely different light. Absent statistical evidence to the contrary, the prosecutor was free to mislead the jury into believing that the odds that appellant just happened to have the same genetic profile as the perpetrator were much rarer: "The coincidences that the defense are going to have you believe in this case are beyond imagination, are beyond reasonable. In order to acquit this man, in order to find him not guilty, you're going to have to look at all of the circumstances of this case, and say 'You know what, he could have been the one in a million.'"

In the respondent's brief to the appeal, the 1 in 3 figure looms large as well, but for the opposite reason. The respondent's brief argued against the validity of the 1 in 3 figure, and the high chances of a jury's accepting it as the probability that Puckett might be innocent of Diana's murder. The brief explains that this understanding is wrong: "The 1 in 3 figure . . . does not come remotely close to conveying the probability that an innocent yet plausible suspect was coincidentally identified in the database search—the only probative question from the jury's perspective."

The basic argument given by the prosecution is that once the match was found, the fact that it corresponded, like the actual murderer, to a heavy-set white California sex offender old enough to have murdered someone in 1972 made it extremely likely that this was not merely a random match to a different person, but the right match.

> In this case, the jury had to determine whether appellant was the perpetrator, or an innocent person who (1) coincidentally shared the perpetrator's DNA profile, (2) coincidentally matched the perpetrator's description, (3) coincidentally lived in and around San Francisco at the time of the crime, and (4) coincidentally committed a number of other violent sex crimes with factual similarities to the assault on Ms. Sylvester. To that end, the jury would have found evidence that the DNA database search produced a "hit" to a plausible but innocent suspect probative. But the "1 in 3" database match probability statistic would not have provided that evidence. It was not, therefore, relevant.

SO WHICH is it? A chance of 1 in 1.1 million of finding a purely coincidental match to a person other than the criminal, or a chance of 1 in 3? The key is to explain the meaning of these figures. Both of them make sense, but neither of them actually gives an estimation of the probability that Puckett is innocent of the crime; they are measuring something different.

The main thing to understand with the 1 in 1.1 million probability is that it has absolutely nothing to do with the chance that Puckett might be innocent. It is simply the probability that a random person has the same DNA as the criminal. Given the US population of about 310 million, this means that one can expect around 300 people around the country to share the same DNA configuration as the sperm sample taken from Diana's body. A priori, then, using no other information about Puckett than his DNA sample, we can deduce that he belongs to a group of about 300 people, each of whom might be the criminal; *thus the 1 in 1.1 million statistic translates into a chance of 299 in 300 that Puckett might be innocent.* In terms of the measure of innocence, this is the *only* information yielded by the DNA statistics, and it is strongly in favor of innocence. The 1 in 1.1 million statistic should not be used alone and out of context, and certainly not as the chance of innocence, although this mistaken impression may have been given to the jury at Puckett's first trial.

As for the 1 in 3 figure, what does it really measure? When calculating that the chance of 1 in 1.1 million of a person's DNA matching the criminal's must be multiplied by 338,000, the size of the database, for a total of 338,000/1,100,000, which is very close to 1 in 3, this calculation is merely finding the probability that there will be a match in the database to the criminal's DNA sample—independently of whether the match really is the criminal or not. Yes, there is a chance of about 1 in 3 of finding a match in the database. But what about the next step? There was only a 1 in 3 chance of finding one, fine; but now one such match has actually been found, and the main question becomes, what is the probability that it is actually the *right* match—to the criminal—as opposed to a purely coincidental match to an innocent man? The answer to this question would indicate the chance of Puckett's guilt or innocence, but this issue is entirely ignored by the 1 in 3 calculation. Like the 1 in 1.1 million, the 1 in 3 probability has a perfectly legitimate meaning—but it is certainly not the chance of Puckett's innocence.

IN HIS article "Rounding up the usual suspects: a legal and logical analysis of DNA database trawling cases," legal scholar David Kaye points out another problem with the 1 in 3 figure. He argues that the presence in the database of a large number of profiles belonging to people either too young to be Diana's murderer or of the wrong ethnic group skews the calculation, since these profiles have no bearing on a search for the murderer. Kaye suggests paring down any database before beginning a trawl, and considering separately the profiles of the *plausible suspects,* those not excluded from suspicion for obvious reasons such as age. If the plausible suspects form, say, 5% or 10% of the database, and the cold hit found in the database is actually a plausible suspect, Kaye suggests, this fact considerably increases the probability that the hit is the right person rather than a purely coincidental match. This reasoning is close to the prosecution's remarks above that argue that the traits shared by Puckett and the murderer (age, race, etc.)—precisely those traits that make him a plausible suspect—should play a role in determining the probability of his guilt or innocence.

If 1 in 1.1 million and 1 in 3 have nothing to do with the probability of Puckett's innocence, then what is that probability? How can we turn the arguments above into a numerical estimation? The purpose of this exercise is not to give an actual mathematical proof; that would be impossible. But a correctly performed estimation is a powerful tool for grasping the nature of a situation. Even though our calculation is approximate, it gives a good idea of the numerical chance of innocence that remains after considering the traits that identify Puckett as a plausible suspect.

We will take the following approach. If Puckett is innocent, then two different people had the same DNA configuration: Puckett and the criminal. Using the reasoning concerning the 1 in 1.1 million figure above, we know that we can expect to find about 300 such people in the United States. These people are expected to be randomly distributed around the country, both by age and geography. Yet Puckett and the criminal share other important traits. They are both white, they were both in California, they are both male sex offenders in the real sense of the word (independently of having ever been caught, registered, or convicted), and they are both over sixty-five, which seems a reasonable minimum age for a person who committed murder in 1972.

To estimate the probability that a random individual in the United States belongs to the group of people sharing these traits with Puckett and the criminal, we begin with the probability of sharing DNA, which is 1 in 1.1 million, equivalent to a probability of about 0.0000009. White people constitute about 72% of the population in the United States, so a person is white with a probability of 0.72. The state of California represents about 12% of the population of the United States, so one is Californian with a probability of 0.12.

The number of registered sex offenders is about 400,000, but it is estimated that the true number is about double that, so let's say 800,000 people. Of these, an estimated 96% are men, which brings us to about 768,000; thus, we estimate the probability of being a male sex offender in the United States as 768,000 in 310 million, or about 0.00247. Considering that people who have been sex offenders at any time in their lives form a group ranging roughly from ages twenty to eighty, those who are over sixty-five would account for a quarter, or 0.25 of this group, so the probability of being a male sex offender over sixty-five is about $0.00247 \times 0.25 = 0.0006175$.

We now have the following table of probabilities:

Probability that a random individual in the United States:

Has the right DNA type:	0.0000009
Is white:	0.72
Is from/in California:	0.12
Is a male sex offender over 65:	0.0006175

The probability of having all of these features, considered as independent of each other, is thus about 0.000000000048. Since the population of the United States is about 310 million, this means the probability that there exists some person in America having all these traits is equal to 310 million times 0.000000000048, or 0.01488, which is roughly equivalent to a chance of 1 in 70.*

At any rate, even though we now know that there is only about 1 chance in 70 that someone exists with all the above traits, we also know that one such person actually did or does exist—namely, Diana's murderer. So the question becomes: what is the probability that a second such person could also exist? Indeed, if Puckett is innocent, he would be the second person—therefore this probability is really an estimation of his innocence.[†]

There is a precise theorem that can be applied to solve this problem: *knowing that one person belongs to a particular group, what is the probability that a second person also belongs to the same group?* Known as Bayes' theorem, it is used quite frequently in legal situations. According to Bayes' theorem, given two events, A and B, the formula for the probability of A knowing B is given by: prob(A given B) = prob(B given A) × prob(A) / prob(B). Here, A is

---

*It should be noted that we have treated race and being a male sex offender over sixty-five as independent here. With accurate statistics we may find that the proportion of whites among male sex offenders of any age is smaller than the proportion of whites in the general population, in which case the product 0.72×0.0006175 would have to be replaced by a smaller figure, which would have the effect of decreasing the 1 in 70 probability.

[†]It is actually slightly larger than the probability of his innocence, since it ignores the tiny possibility that two different people could belong to the group and yet Puckett be the murderer rather than the second person.

the event "two people belong to the group," and B is the event "one person belongs to the group." Our question can be rephrased in mathematical language as: *knowing B, what is the probability of A?* We know that the probability of B is 1/70, and the probability of A is thus $(1/70)^2 = 1/4,900$. As for the probability of B given A, that is equal to 1 (in other words, certainty), since if A is true (two people belong to the group), then B is certainly true (one person belongs to the group). So prob(B given A) = 1; prob(A) = 1/4,900; and prob(B) = 1/70; thus Bayes' theorem tells us that prob(A given B) is equal to 1/70. In other words, in this situation the probability of a second person belonging to the group, knowing that one person already belongs, is exactly equal to the probability that one person belongs.

From these considerations, we conclude that there is at most a chance of about 1 in 70 that John Puckett could be a different person from the one who left the sperm sample at the murder scene. Unlike the figures discussed by the defense and the prosecution, this is a legitimate estimation, albeit an approximate one, of the chance of Puckett's innocence.

John Puckett may be guilty or he may not; this book is not a tribunal. What is certain is that he must not be either convicted or acquitted based on a calculation of something quite different from the probability of his innocence. If mathematics is to play a role in his trial, it should—it *must*—be correct mathematics. Anything else is a travesty of justice.

MATH ERROR NUMBER 6 »
SIMPSON'S PARADOX

The average SAT verbal test score in 2002 was precisely the same as it was in 1981. Yet each of the six major ethnic categories used by the College Board shows an increase in scores over that period of time: whites, 8 points; blacks, 19; Asians 27; Puerto Ricans 18; and American Indians, 8. How can it be, then, that all groups that make up the national average have gained but the national average score has not budged in 21 years?*

THIS PUZZLING FACT is a classic example of the phenomenon known as Simpson's Paradox: over the last twenty years, the average score of every ethnic group of students on a standardized test administered every year has gone up—yet the overall average is identical to what it used to be. Table 6.1 shows the average test scores of each group in 1981 and in 2002, with the overall averages at the end. *Every group has improved, yet the overall score is identical.* How can that be?

The secret lies in the important role played by a factor not displayed in table 6.1. In the present case, that factor is the large-scale changes that occurred within the populations of the different ethnic groups with respect to one another. In particular, as table 6.2 shows, the size of all the minority

---

*From EDDRA, the Education Disinformation Detection and Reporting Agency.

TABLE 6.1

	White	Black	Asian	Mexican	Puerto Rican	American Indian	Total
1981	519	412	474	438	437	471	504
2002	527	431	501	446	455	504	504

TABLE 6.2

	White	Black	Asian	Mexican	Puerto Rican	American Indian	Total
1981	85%	9%	3%	2%	1%	0%	100%
2002	65%	11%	10%	4%	1%	1%	92%*

*The remaining 8% of the population belongs to other minorities.

groups increased significantly in proportion to the white population. This table shows the ethnic makeup of the test-taking populations in the two years under comparison.

What table 6.2 reveals is that while the white group has the highest average score on the test in both 1981 and 2002, in 2002 whites count for much less overall in the calculation of the average score. The lower scores of the other groups bring the average down, even while the increase in the average score of each group brings the average up. The overall result is: no change whatsoever.

Because of these invisible factors, Simpson's Paradox can easily be manipulated to deceive. As an employee of a major gas company told a mathematician friend of ours, "I had to present the economic results of the company over the last year, so I asked them whether they wanted a presentation that makes it look like we earned money, or one that makes it look like we lost money."

Simpson's Paradox never fails to astonish whenever it shows up in real-life situations, and it serves as a constant reminder that conclusions drawn from statistics are a lot less clear-cut than we think.

The Berkeley Sex Bias Case:
Discrimination Detection

It is one of the most prestigious universities on the planet. The Academic Ranking of World Universities has rated it third-best school in the world for years, squeezed in among Harvard, Stanford, and Britain's Cambridge. Students dream of admittance; more than half of the applicants—not to mention the successful ones—have perfect grades in high school and stellar results on their standardized tests. The campus crawls with vigorous young people intent on success. And to top it off, the university is located in the historic town of Berkeley, the "third most politically liberal city in the United States," one of the birthplaces of the 1960s hippie movement, and among the best places in America to sip a mocha latte while stretching one's legs at an outdoor café. Lush gardens surround the elegant campus buildings, and long palm fronds wave overhead against a shimmering, flawlessly blue sky; to many, the University of California at Berkeley may seem like the Promised Land.

The University of California at Berkeley

But Berkeley is by no means a flawless institution. It has undergone periods of public criticism, and its high profile has ensured that the university has been widely covered in the press and deeply investigated within the school itself. One of the recurring problems the university has faced is allegations of sex bias: giving unfair advantage to men over women, both in student admittance and in faculty hiring.

JENNY HARRISON probably knows more about the consequences of this type of allegation than any other person alive. In 1975 she earned a PhD in mathematics from the United Kingdom's University of Warwick. The research she published there was recognized by fellow mathematicians as being brilliant and quickly secured her prestigious postdoctoral positions at Princeton and then Berkeley, where she took up a tenure-track position in 1978. At that point in the young woman's career, everything looked ripe with promise.

Jenny continued with her research at Berkeley, but she encountered a few problems along the way—her work was difficult to understand and aroused disagreement among her colleagues as to its correctness and its value. These disputes were eventually resolved sufficiently for her to publish a paper in 1986, eight years into her position. That was the same year when her case came up for a tenure decision by the department. Jenny felt confident going into the tenure review: her work was now going well, and several of her colleagues had already received tenure since she had first arrived there.

But she was in for a shock. The mathematics department, citing the standard of her work as not high enough for a permanent professor, denied her tenure. It was the first time in twenty years that the math department at the university had not granted tenure to one of its own.

JENNY HARRISON was immediately suspicious. She knew the other assistant professors who had been granted tenure since she had been at Berkeley, and she didn't consider their work to be markedly superior to hers. Jenny became convinced that there was another cause—namely, that some influential members of the department objected to her because she was a woman. Indeed, a quick look at the department lent some weight to her claim. The tenured faculty in mathematics numbered only a single woman among its sixty-odd members.

With the support of a few department members who agreed with her, Jenny filed a complaint with the Privilege and Tenure Committee, a campus-wide group of professors who reviewed grievances regarding tenure cases. She alleged that her file had been unfairly judged and that the unfairness was due to gender bias. The committee looked into the situation and ultimately ruled against her. But Jenny felt she had not been given a fair chance to present her case. She had been denied access to the documents she would have needed to prove her point—most importantly, the files of other recently tenured department members. Even her own file had been partially withheld. In contrast, the university's staff had had complete access to everything. It was simply not a level playing field.

So Jenny went to court. She filed a lawsuit against the University of California at Berkeley for gender discrimination. With this act, closely guarded secrets of the legendary mathematics department began trickling out into the open. One after another, professors involved in the dispute were questioned by lawyers. Hours and hours of such questioning produced hundreds of pages of documents, and details began to emerge.

One of the most commonly cited arguments against giving Jenny tenure was that there were other mathematicians "out there"—stars, Fields Medalists, etc.—whose work was more brilliant, more exciting, or more revolutionary than hers, and who could be offered tenured positions instead of her. Jenny countered that this wasn't valid reasoning; otherwise it should be applied to every member of the department who came up for tenure, in which case almost no one would receive it. After all, there is practically always someone more brilliant out there. Being measured directly against the world's very greatest mathematicians, she claimed, was a process that had been applied to her alone: the only woman candidate. The proper assessment would have been to compare her with other assistant professors of mathematics at Berkeley who had come up for tenure and to whom it had been granted. (It is notable, and may perhaps also be a sign of sex discrimination, that the only other female faculty member at Berkeley, who had been granted tenure in 1975, was considered significantly above this group in excellence.)

Between 1978, when Jenny first took up her position, and 1988, when the Privilege and Tenure Committee reviewed and rejected her grievance, there had been no fewer than eight other applications for tenure in the mathematics department, all of them successful. At first Jenny had not been allowed to see these other applicant files in order to publicly compare them with her own, but once the lawsuit was initiated, she was able to obtain a court order obliging the university to deliver them to her. Of course, no two mathematicians will ever agree completely about exactly whose work is more interesting or better than whose; to some extent it is a subjective matter. But according to at least some of her colleagues, the files indicated that Jenny sat squarely in the middle of the overall group; there was nothing in her file that made it significantly worse than those of the other tenure applicants. The comparison lent weight to her charge of gender discrimination.

Jenny also found more tangible traces of discrimination. She discovered, for example, that in sending out a letter requesting an opinion on her suitability for tenure, the department chairman had thought fit to add that the evaluator should keep in mind that Berkeley aspired to be "the top mathematical center in the world." This phrase was used in soliciting opinions about Jenny, but it had not been used in the solicitation letters concerning any of the other candidates.

Rather than go to trial, the university finally agreed to a settlement with Jenny: a committee of reviewers from outside the university would be called in to judge the situation. As weeks dragged into months and months into years, however, one decision was made that worked in Jenny's favor: the outside committee decided that it would consider not only her work from 1978 to her tenure rejection in 1986 but her subsequent work as well, and they would use it all as a basis to judge whether she was good enough to take her place among the other tenured professors at Berkeley. During the seven-year battle, Jenny bore and raised a child, overcame a bout of throat cancer, and produced enough new research so that when the decision finally came in 1993 it was in her favor. Jenny Harrison was awarded a financial payment (amount undisclosed), and above all she was finally appointed to a tenured position at Berkeley. The numbers have increased a bit since then; today, Harrison is one of four female full professors in the Berkeley math department, alongside about fifty men.

The final settlement was not to everyone's liking. Loudest in their disapproval were the professors who had been accused of sex bias in denying Jenny tenure. Needless to say, no single individual ever admitted anything of the kind. It is a difficult accusation to prove—after all, who can penetrate into the mind of another and know what he or she is thinking? There were professors in the department who scoffed at the idea of sexism playing any role in their vote, and there were others who were more reflective on the matter. In the case of a male candidate, observed one of the latter, "there's hardly ever an important split in the department," but "if a woman is perceived as being 'not world class,'" then "all these doubts come out" and she is treated differently from men with similar qualifications.

Jenny Harrison, professor of mathematics at Berkeley

But even if Jenny successfully proved that she had been treated differently from the other candidates, how could she possibly show that that difference was due to sex discrimination? After all, there could be any number of other factors at play that no one wanted to mention: incompatibilities, rivalries, jealousies, and just plain mathematical judgments, which must always contain an element of subjectivism. If no one ever admits to negative feelings about women, whether as mathematicians, colleagues, professors, students, or in general, then how can anyone ever hope to actually prove that those feelings exist?

The only way to prove or disprove such a claim, in fact, is to leave the individual case aside and focus on trends. If one can point to the existence of a regular, repeated pattern of behavior, then one may begin to gather elements of actual proof.

DURING THE course of the Jenny Harrison tenure battle, faculty members, both men and women, publicly opined on the matter. Professor Lenore Blum, who was not a faculty member at Berkeley but held the position of deputy director of the Mathematical Sciences Research Institute, the MSRI, located just up the hill from the university campus, stated quite emphatically that problems did exist. "Blanket denials that [the department] ever behaved badly towards women—or categorical statements that its actions have been

exemplary—are just not true, nor are they widely believed outside Berkeley,"
she said. And she went on to describe the university's reaction to the pressure exercised on all American public institutions in the 1970s to make a
special effort to recruit both women and minorities. Obeying the letter, if
not the spirit, of the affirmative action trend, Berkeley issued an invitation
for women and minorities to apply for two tenure-track positions that had
opened in the department. But behind the scenes, "the Department had already offered the positions to two men," claimed Blum. "This charade was
clearly unfair to the women and minorities who applied in good faith and
were subsequently subjected to an evaluation which necessarily had to unearth flaws in their records." In other words, in order to reject all the candidates who had been so warmly invited to apply, the hiring committee had to
search out bad things to say about them in order to explain the rejection.
This didn't exactly cast the university in the best light.

And Berkeley's record with actual hiring is certainly no proof that sex
discrimination did not exist. When Marina Ratner—the only female tenured
faculty member in the department during the Harrison case—was first hired,
a letter was printed in the campus newspaper written by a department member who stated that although Ratner was "well-qualified," there were "several
men who were better." This attack seems gratuitous, coming as it did *after*
the department had already voted to offer the woman a position. The goal
appears to have been purely to denigrate her, with no practical purpose in
mind. Of course, as above, it is open to doubt whether this sentiment was
or was not driven by sexism; the author of the letter might have written exactly the same letter for a newly hired male with the same qualifications.
Whatever the case, his assessment of Ratner was simply wrong. She went
on to become one of Berkeley's most famous mathematicians, proving astonishing theorems, winning major prizes, and being elected to the National
Academy of Sciences. But Ratner herself does not view the letter as being
motivated by sexism, adding that although she believes sexism does exist in
the mathematical community, "it is mostly subconscious."

This is possible, even probable. The men who examine the women's applications and find things to denigrate, criticize, and belittle are probably not
consciously aware of any motivation for doing so other than a lofty desire to
make a correct judgment. They are undoubtedly acting in good faith, if not
always with elegance. Certainly in Jenny Harrison's case, the outside review

committee who eventually recommended her appointment with tenure compared her file with those of the other tenured professors and found no significant difference in level. But that was several years after the first tenure review, and she had much more published work the second time around. Some in the department assert that her rejection in 1986 was a judgment based purely on mathematical considerations, while others claim that her mathematical level was already sufficient at that time and there were other causes. The conclusion is that in an individual case, it is virtually impossible to ascertain the precise causes of an event such as tenure rejection, and the precise role that sex discrimination might have played. The existence of sex discrimination cannot be detected by staring into the minds of individuals. It is necessary to step away from the personal stories, and consider the larger trends.

IT SO happened that shortly before Jenny Harrison first arrived at Berkeley, the university had been sued for bias against female applicants to graduate school. Indeed, a simple examination of the overall data showed that of the nearly 13,000 applications for doctoral spots at the university—broken down into about 8,500 men and 4,500 women—fully 44% of the men had been accepted, and only 35% of the women. The numbers are given in table 6.3.

	Number of Applicants	*Number Admitted*	*Percentage Admitted*
Men	8,442	3,738	44
Women	4,321	1,494	35

TABLE 6.3

If not sex bias—if all judgments about whom to accept had been made with total objectivity on the basis of the quality of the applicants—then one might conclude that the men sent in higher-quality applications than the women, or perhaps that men are better students in general. But all data strongly belie this assumption; women generally outperform men at the undergraduate level in the United States, so if anything their applications would have been stronger than the men's. Why, then, were so few of them, relative to the male candidates, considered worthy of acceptance to the ivory tower?

When these results were published, it seemed to many that here, finally, was clear proof of the existence of sex discrimination at Berkeley.

TABLE 6.3 represents admissions only from the academic year 1973–1974. Before launching accusations, it is important to determine whether these numbers could represent a normal statistical fluctuation, something that might happen quite naturally once or twice over a significant period of time.

To make such a determination, we must calculate the chance of such numbers—44% versus 35%—happening naturally, under the assumption that there is no sex bias or inherent inequality in the applications, so that women and men have the same a priori chances of getting in. If this chance turns out to be, say, 1 in 20, we could conclude that once every twenty years or so the admissions figures can be expected to look like this, and so it may not really be a suspicious circumstance indicating the existence of bias.

The total number of applicants is 12,763, and the total number of admissions is 5,232, which is 41%. So under the hypothesis of complete equality, one would expect 41% of men and 41% of women to be admitted, or 3,461 men and 1,771 women.

TABLE 6.4

	Number of Applicants	Number Admitted	Number Expected
Men	8,442	3,738	3,461
Women	4,321	1,494	1,771

From the figures in table 6.4 we see that in reality Berkeley accepted 277 more men and 277 fewer women than expected. The number that gives the probability of such a skewed result happening naturally in a pool where everyone has equal chances is called the *p-value*.* Under the assumption that men and women candidates are of equal level and stand an equal chance of being accepted, the p-value measures the probability that this particular distribution would occur over many (theoretical) years of admissions.

For a simple explanation of the meaning of the p-value, suppose you have a vat containing 8,442 black marbles and 4,321 white marbles (the number

---

*This p-value will play an important role in the Lucia de Berk case (see chapter 7).

of male and female candidates), and without looking at their colors, you randomly select 3,738 + 1,494 = 5,232 marbles (the number of accepted students). The p-value is the probability that you will select at least 3,738 black marbles and at most 1,494 white ones. It is easy to run a simulation of the experiment millions of times on a computer, and there is also a theoretical formula for the p-value. Once calculated, it turns out that the p-value in the Berkeley case is extremely tiny, equal to 0.0000000057, or about 6 chances in a billion.

Generally, when the p-value for an occurrence is under 1 in 1,000, it is accepted that there is legitimate cause to ask whether the hypothesis of pure chance may be wrong and to investigate whether there is another cause. One has to leave room for the surprise factor, though, because very unlikely events (sometimes called "black swans") do constantly occur in the world. The bottom line is that unlikeliness and rarity are legitimate grounds for suspicion but are not sufficient to draw any conclusions. All that can be said is that there are grounds for further analysis. And that, rightly, is the view that the University of California at Berkeley decided to adopt.

THE UNIVERSITY called together a committee of three to perform the investigation: a professor of statistics; a professor of anthropology, who was also dean of graduate studies; and one member of the data processing staff of the graduate division. The three committee members began by deciding that since admissions at Berkeley are decided department by department, independently of each other, it was possible that the cause of the inequality was due to a lack of evenhandedness in just a few departments rather than a single cause across all departments. Therefore, the committee requested a breakdown of admissions data by department, and began their review of the situation by exonerating those departments that either had not received any applications from women at all or had accepted all students who applied.

After this exercise, eighty-five departments remained under scrutiny. The committee examined their admissions statistics one by one, and in each case the p-value was calculated as above to see whether the fluctuations that occurred in the percentages of men and women being accepted were really out of line with numbers that might occur with a reasonable frequency in the absence of any bias. Most of the departments showed no unexpected gap in the percentages of men and women accepted from the applicant pool. In the end, the committee narrowed the departments under question to six. These were six large departments, each of them with numerous applicants. The admissions data of these six departments lumped together are given in table 6.5.

	TABLE 6.5		
	Number of Applicants	*Number Admitted*	*Percentage Admitted*
Men	2,590	1,192	46
Women	1,835	557	30

It would seem that the culprits had been correctly identified—nearly half the men had been accepted, and less than one-third of the women! The ratios were even worse than the 44% versus 35% of the university overall. Clearly, it would seem, these departments housed the university's most misogynous minds, shunting Jill's application into the wastebasket while placing Jack's in the acceptance pile.

The investigators went further, deeming that the worst of the offenders should be identified, the finger correctly pointed, and the situation rectified. To this end, they made a table listing the acceptance figures for the six departments separately (table 6.6).

	TABLE 6.6					
Department	*Number of Male Applicants*	*Number of Male Admissions*	*Number of Female Applicants*	*Number of Female Admissions*	*Pencentage of Male Admissions*	*Percentage of Female Admissions*
A	825	511	108	89	62	82
B	560	353	25	17	63	68
C	325	120	593	202	37	34
D	417	139	375	131	33	35
E	191	53	393	94	28	24
F	272	16	341	24	6	7

But to their surprise, this more detailed table didn't give any indication of bias against women at all! Out of the six departments, four of them (A, B, D, and F) actually accepted *higher* proportions of women than of men, with department A even showing a marked preference for the female candidates, accepting 82% of all female applicants compared to only 62% of the males.

The remaining two departments, C and E, showed only a tiny edge for the male candidates: 37% of males versus 34% of females for department C, and 28% versus 24% for department E. Certainly nothing to raise an eyebrow at. Where did that elusive sex bias go? It seems to have melted away, and we are left with a paradox. No single department shows marked bias toward male candidates, with most actually giving an advantage to females. Yet the overall totals reveal that a far lower percentage of women than of men are accepted. How can that be?

THIS PROBLEM, which commonly occurs in statistics, is known as Simpson's Paradox, and it arises through the act of forgetting, or ignoring, an important piece of relevant data. Here that piece of information is the following: what proportion of women and what proportion of men actually apply to the departments with the lowest (or highest) acceptance rates?

To show how tricky Simpson's Paradox can be in real-life situations, let's drastically simplify the problem and suppose that there are only two departments in the university, A and B. Suppose there were 1,000 candidates in all, 600 men and 400 women, and suppose that both departments actually have quite a strong bias *in favor* of the female candidates. But suppose also that acceptance rates at B are overall much lower than those at A—and that most of the women apply to B! Then we might have a table like table 6.7 on the next page.

This simplified situation reveals in a nutshell what was happening at Berkeley. Far from showing bias against women, department A accepts 90% of women compared to only 80% of men, and department B accepts 30% of women compared to only 20% of men. Yet the total acceptance rates show that 70% of men are accepted compared to only 45% of women!

It's enough to make anyone suspicious, and yet the problem doesn't even exist. Or at least there is a problem, but it is absolutely not about sex bias in the admissions process.

IN THE END, the skewed admissions data at Berkeley didn't reveal any sex discrimination in the admissions process, but it did reveal another factor, something that was already well-known. So well-known, in fact, that it didn't occur to anyone that it was actually causing the problem in acceptance figures namely, the fact that very few women apply to the Berkeley mathematics and engineering departments. Few apply, and even fewer are accepted,

TABLE 6.7

DEPARTMENT A

	Number of Applicants	Number Admitted	Percentage Admitted
Men	500	400	80
Women	100	90	90

DEPARTMENT B

	Number of Applicants	Number Admitted	Percentage Admitted
Men	100	20	20
Women	300	90	30

DEPARTMENTS A AND B TOGETHER

	Number of Applicants	Number Admitted	Percentage Admitted
Men	600	420	70
Women	400	180	45

whether as students or as faculty. The two separate problems—the department's record in rejecting women and the lack of women applicants—are actually closely related.

Why do so few girls choose to study math and engineering? That question has occupied many a student of society, and many different factors have been blamed, ranging from the psychology of the individual, to deeply rooted social attitudes, to differences of physiology and child-rearing methods. While no satisfactory answer has ever been given, it is generally accepted that one of the factors that discourage young women from studying mathematics is the perception of the subject as being male-dominated and competitive. This is a vicious circle, since the lack of women within mathematics departments really does perpetuate the male domination of the subject, making it a hard-to-breach fortress for female candidates. That fortress was precisely what Jenny Harrison had to face and vanquish. As it turned out, the whole question of sex discrimination at Berkeley was really a reflection of a much wider problem—the problem of gender and mathematics.

MATH ERROR NUMBER 7 »
THE INCREDIBLE COINCIDENCE

SUPPOSE YOU BUY a lottery ticket. You wake up the next morning and are shocked to find out that you have the winning number. Certainly that is an event with a very low probability; yet at the same time, someone is certain to win the lottery, so that "coincidence" is sure to occur somewhere, for someone. The only difference between you and that other someone lies in your point of view. Statistically speaking, the two events—"I win" and "He wins"—are equally probable at the start.

On the one hand, we all know that for each lottery, some John Doe will win. But at the same time, when John Doe does win (especially if he happens to be you), you realize that the probability of that happening was extremely tiny. This may sound like a contradiction, but of course it isn't. It's merely an illusion arising from the fact that until John Doe won, you made no specific prediction concerning him, whereas when you bought your own ticket, you probably thought something along the lines of "I have one chance in a million of hitting the jackpot."

Because of this, it can be misleading to retroactively calculate the probability of an event that has already occurred. If after performing such a calculation you find the probability of the event's occurring to be extremely small, you might become suspicious and wonder if whatever it is that you observed happened as a result of scheming rather than chance.

In the event of a crime, such calculations can become slippery—one must take enormous care. While it might be legitimate to have suspicions, they might be the result of this kind of retroactive thinking—and when the police act on a suspicion, lives can be destroyed.

The Case of Lucia de Berk: Carer or Killer?

On the morning of September 4, 2001, at Juliana Children's Hospital in The Hague, a baby died unexpectedly. Little Amber was almost six months old and had struggled since birth with a complicated condition involving anomalies of the heart, brain, lungs, and intestine. She was not able to eat on her own and had to be fed artificially. On July 25, she had had a heart operation that seemed to help, but after a month of improvement, Amber suddenly needed extra supplies of oxygen and diuretics on August 28, and her condition worsened. By September 3, Amber was suffering from vomiting and diarrhea and appeared to be in pain.

Two nurses were in charge of Amber's care. One of them was Lucia de Berk, a forty-year-old licensed pediatric nurse from The Hague. At around 11:00 p.m. on September 3, Lucia decided to connect Amber to a monitor to keep close track of her heart rate and breathing difficulties. She also called for doctors to come and examine the baby, who seemed to be growing steadily worse. Amber was wheeled to the examining room, and two pediatricians examined her; the time of the examination was recorded as 1:00 a.m. on September 4. They put her on a drip and diagnosed her with enteritis, an inflammation of the small intestine, but they did not judge the child to be dangerously ill. After the examination, they sent Amber back to her room, where she was reconnected to her monitor by finger cuff.

At 2:46 a.m., little Amber went into crisis. To the horror of the two nurses in the room, the baby's breathing frequency dropped suddenly and drastically, followed by a slowing of her heartbeat. Her face turned gray. They called for a doctor at once, and he immediately summoned the resuscitation team, but it was impossible to save the dying child. They spent forty-five minutes trying to revive her, but she was declared dead at 3:35 a.m. Her heart had actually stopped beating sometime earlier.

The doctors familiar with Amber's situation did not find her death suspicious; they knew how ill she had been. But the doctors who were present that night were not the ones involved in her regular treatment. Nevertheless, they signed a declaration of natural death.

By the next afternoon, however, that declaration had been officially retracted by the hospital.

ON THE day after the death, a nurse at the hospital heard about the incident and went to talk to her superior. As she explained it, she was "worried that, during her 2 years at Juliana Children's Hospital, Lucia had been present at five resuscitations." It seemed to the nurse that this was a large number compared to the experiences of other nurses. Her superior agreed, and a rumor began to make the rounds: a list of the five resuscitations where Lucia de Berk had been present was soon circulating around the pediatric ward. Looking at it, the other nurses had to agree that it seemed to be too many to be attributable to simple coincidence. Worse, Lucia had been present at five patient deaths.

The categorization of Amber's death was changed to "unnatural," and after what must have been a heart-wrenching conference, the situation was brought to the attention of the general director of the hospital, Paul Smits. Smits was the director of two hospitals in The Hague: the Red Cross Hospital and Juliana Children's Hospital. He had some expertise in making Microsoft Excel spreadsheets, and together with the chief pediatrician he proceeded to do a little computation of his own. Putting together all the information the nurses had brought him, he felt he ought to calculate an actual figure for the probability of Lucia being present at so many resuscitations and deaths. What he found disturbed him deeply. Even if his calculation was not exact, it seemed to him, as it had to the nurses, to indicate that "Lucia was involved in an extremely unlikely, high number of incidents."

On the morning of September 5, the day after Amber's death, the five other deaths at which Lucia had been present were all reclassified as "unnatural"; they had all been declared natural when they occurred. To collect the largest amount of data possible, Mr. Smits contacted his other hospital as well as the Leyenburg Hospital, where Lucia had previously worked, and asked for the list of deaths at which she had been present. When the lists arrived, they seemed to bear out his fears. The doctors, who were now very concerned, pushed him to take the affair as seriously as possible. Mr. Smits called the police and Lucia de Berk was formally accused of thirteen murders and four attempted murders.

A DIRECTOR of two hospitals beset with financial and organizational difficulties, Paul Smits attacked the problem as he did the innumerable problems that faced him each and every day: with authoritarian vigor. It was clearly a situation in which action had to be taken swiftly and efficiently. That meant not only removing Lucia as quickly as possible from the active nursing staff or handing the case over to the police, but also informing the media of what had occurred.

Smits took all the information from the three hospitals and turned it over to the police. He claimed that he did not give them his calculations or their results; those numbers simply influenced his decision to call them in. However, it is clear from records of the police interrogations of nurses in those first days that they were already working on probability figures.

Next, the board of directors of the hospital contacted the Netherlands' most sensational newspaper, *De Telegraaf*. The article that soon appeared in *De Telegraaf* did not name Lucia, but it told the terrifying story of a nurse in two hospitals who had been involved in the killings of large numbers of patients. The newspaper transmitted the director's sincere apologies, his sympathy with the families of the victims, and his desire to get to the bottom of each and every incident and to make sure that punishment was meted out where it was due. The article treated the deaths as murders, although they all had been declared natural deaths until it had been noticed that they happened on Lucia's watch. To top it all off, the paper hinted that more murders would probably be surfacing soon as the investigation continued. Within a single day, that publication turned Lucia de Berk into the most horrific serial killer the Netherlands had ever seen.

Other papers picked up the story. Lucia was compared to Beverley Allitt, baptized "The Angel of Death" (see chapter 1); this epithet was even used in court. The probability figures that the police had already considered leaked to the papers, with figures such as "a single chance in seven billion" bandied about— a devastating number, given that there are only six billion people in the world. It goes without saying that no mathematical justification of that number was ever included in the articles where it appeared. By unstated consensus, numbers in newspapers carry their own justification, or at least their own prestige, along with them.

Lucia de Berk

It was not enough for Lucia to be banned from her job and her livelihood, and to be under suspicion of unspeakable crimes. She was painted as the most vicious creature that the public imagination could conceive of—a slaughterer of babies and elderly people, a destroyer of the weak, a vile monster. She had no way to defend herself other than to repeatedly declare her innocence.

On December 13, 2001, Lucia de Berk was arrested and charged with thirteen murders and four attempted murders. Even though it cannot have come as a surprise, she was stunned, stating that she had no knowledge of any of the acts attributed to her. She was remanded in custody while awaiting trial.

LUCIA'S FIRST trial began several months after her arrest. It emerged almost immediately that there were a few peculiarities about the accusations leveled against her. To start with, it was soon proven by her lawyers that there were two deaths on the list contained in the accusation at which Lucia had not been present at all. Either she had signed out and left the hospital before they occurred, or she had been out sick at a time that normally would have been her shift. There was also a case or two in which the death was so obviously expected and natural that they could not reasonably form the basis of any accusation of wrongdoing. Those cases, however, were quietly dropped from the trial. No one tried to calculate the differences they made to the damning "probabilities."

Next, it was set forth by Lucia's lawyers that not a single one of the deaths, or even the nonfatal incidents included in the list of accusations, had been observed to be in any way unnatural when it occurred. They emphasized the peculiarity of the situation: until the day when someone noticed that Lucia was present at a surprisingly large number of these events, there had been no grounds for suspicion of anything, of any kind, at all. In response, the prosecution argued that Lucia must have covered her tracks extremely well.

Finally, and this was perhaps the most difficult point, even once the deaths had been accepted as murders, there was no medical evidence to justify that claim. No traces of poison or violence could be found in or on the bodies that had been exhumed for the purpose,* and the medical witnesses called to the stand found it hard to show why they believed the deaths were due to murder at all. It was not as though such a situation was legally unknown—it had already occurred for the mothers accused of crib death, for example—but it was still uncomfortable. The prosecution made an extra effort to locate evidence of murder, sending for repeated medical analysis every bit of the physical remains of the long-dead victims that they could lay hands on.

In the case of little Amber, a jar in the hospital storehouse contained gauzes soaked with diluted bloody fluid from her body taken at the time of her autopsy. Laboratory tests run on this fluid produced evidence of a small but lethal concentration of a drug called digoxin. The doctors claimed there should have been no trace of digoxin in Amber's body, because even though she had been given the drug during the first four months of her short life, the treatment had been stopped two months before her death and all traces would have been expelled by then.

Evidence was presented at the trial that Lucia had a key to the hospital cupboard where the digoxin was kept. In the absence of any further confirmation of the poisoning, another suspicious fact was revealed. Hospital records showed that about an hour and a half before Amber died—about

---

*It was said by many that Lucia had purposely chosen Islamic patients as her victims knowing that the parents would not accept autopsy or exhumation of the bodies. After two years, however, many of these families ended up legally obliged to accept them in spite of their religious beliefs.

the time lapse it would take for a digoxin injection to kill her—the baby had been disconnected from her monitor for a period of about twenty minutes. These two facts—the key and the disconnection—were cited as Lucia's *means and opportunity* to murder Amber.

The prosecution managed to unearth some evidence indicating a possibility of poisoning in a second case: that of a boy named Achmad, a very sick child who had died at Juliana Children's Hospital several months before Amber. On January 25, 2001, Achmad had gone into a coma from an overdose of chloral hydrate. The medicine had been prescribed by the neurologist for him in rather large quantities in case of restlessness, but once Lucia had been accused of murder, it became easy to conclude that she had deliberately administered an overdose. Thus means and opportunity were once again present. Achmad's coma was one of the nonfatal incidents of which Lucia was accused. His death a month later, due to an error in his medicine, was one of the accusations of murder.

Achmad underwent a gastroscopy, for which he was given anesthesia on the morning of February 23. That same evening, he was prescribed two sedatives: dipiperon and oxazepam. The doctor who prescribed these had not intended him to receive a large dose of chloral hydrate as well, but as no one removed it from the ward dossier, it was given to him in addition to the other medicines. Although Lucia had already left for the night when the poor child succumbed to the combined effects of serious illness, the shock of operation, and the effects of overmedication, it was assumed that she had administered the medicine before leaving. After consultation with the coroner, Achmad's death was declared natural. But in hindsight, it seemed only too easy to assume that Lucia had deliberately increased the dose of chloral hydrate—already large for his size, age, and condition—to an overdose inducing death.

Of all the other murders attributed to Lucia, not a single one could be attributed specifically to any drug. Thus, the cases of Amber and Achmad became known as *locomotive cases,* meaning that if it could be proven that Lucia had committed these two murders, then it could be inferred that she must also have committed the others, dragged along behind the locomotives like the wagons of a train. This is known as juridical chain evidence.

The theory, such as it was, made at least some sense, but there was a disturbing lack of medical proof. Investigators began to hunt for other types of evidence, clues that could reveal what was behind Lucia's behavior. For this purpose, they seized her private diary and found a suspicious and frightening entry for November 27, 1997. There, Lucia had written: "Today, I gave in to my compulsion."

It so happened that November 27, 1997, coincided with the death of a Mrs. Zonneveld, an elderly patient of Lucia's in the terminal stage of cancer. The prosecution called Mrs. Zonneveld's surgeon as a witness. Of course he had certified the woman's death as natural when it occurred, but after the trial, he stated in a letter to the court that he had been rather surprised when she died, as he had expected her to live a few days longer.

When interrogated about her "compulsion," Lucia explained that she was referring to her addiction to reading tarot cards for her friends and family, as well as for her patients—a practice strongly frowned upon in hospitals. Lucia owned a pack of tarot cards, which she kept carefully wrapped in a traditional wooden box, and her psychologist stated that her tendency to resort to the cards for insight, and to yield to that tendency as a compulsion, corresponded to her personality. But the idea was laughed out of court. It was all too trivial. It was so much more meaningful to assume that Lucia had yielded to an unspeakable compulsion to murder her dying patients.

Digging for dirt into her past, the investigators discovered a brief involvement in prostitution at the age of seventeen, when Lucia had been living in Canada. They also learned that once in the Netherlands, she had falsified a Canadian school diploma in order to enter nursing school. She had no morals, it was concluded; she was a liar and a cheat. Little by little a picture of a murderer was building up. It hung together—yet almost none of it constituted real proof.

The real clinching factor, the one that had convinced the hospital's director in the first place, was not Amber's death, or Achmad's, or even the completely unexpected resuscitation of a little boy named Achraf on September 1,* which was later recognized by many as the first real moment

---

*One-and-a-half-year-old Achraf had been admitted to the hospital as a "social case"; his mother was worried about his apnea, but he was not perceived as seriously ill. No one realized for some time that the child was suffering from a severe hereditary illness, Freeman Sheldon syndrome, with abnormalities of the heart and lungs. The necessity for his resuscitation took doctors by surprise and was an important catalyst in starting the accusations against Lucia.

when they realized that something seemed to be wrong with Lucia. Nor was it a matter of diary entries or youthful misbehavior. Instead, it was the statistical analysis that finally persuaded Smits: the table that he and the doctors had made showing the proportion of deaths and dangerous incidents that had occurred during Lucia's shifts.

As the trial developed, this table became the most damning item: the one thing people could not dismiss by saying it was either irrelevant, unimportant, or not proven. But to make proper use of the table, it was not enough to show the raw numbers to the jury, or to give them the intuitive calculations that Smits had made with the help of the doctors now pushing for Lucia's arrest. The conclusions of a professional statistician were required. To calculate the real probability that the numbers in the tables could indicate murder was a task that should have been entrusted to an expert.

IT WAS unfortunate, then, that the main expert witness the court chose to call to analyze the table of numbers was actually a law professor with an undergraduate degree in statistics. The Netherlands is home to any number of internationally renowned professors in mathematical statistics, but the expert called to the trial was Henk Elffers, a professor of law and psychology specializing in the psychology of compliance and spatial crime analysis. Elffers used the knowledge he had gained as a young student to attack the particularly difficult problem of decoding the meaning of the information contained in Smits' murder table. He took the table of deaths and serious incidents given to him by the hospital doctors, used it to make his computation, and gave the result to the court. Elffers' conclusion: there was one chance in 342 million that Lucia could coincidentally have been present at so many natural deaths.

Although neither the table nor the computation was fully correct, as we shall see, nevertheless, when one considers the evidence, it is not hard to understand why Paul Smits became suspicious and why Henk Elffers believed it indicated guilt with virtual certainty. During the nine months that Lucia worked at Juliana Children's Hospital, 1,029 different nursing shifts had taken place, and she was present on 142 of them. Eight of the incidents during Lucia's shifts that the hospital had reclassified as "unnatural" occurred during this nine-month period. The data collected by the doctors and nurses at the Juliana hospital is given in table 7.1.

TABLE 7.1

Juliana Children's Hospital Shifts	Number of Shifts without Incident	Number of Shifts with Incident	Total
Without Lucia	887	0	887
With Lucia	134	8	142
Total	1,021	8	1,029

Admittedly, the numbers are striking, even worrisome. There is no doubt that Lucia was present at far more near-lethal incidents than one might expect of a random distribution of nursing shifts.

Fortunately, statistical analysis exists in order to give a precise mathematical estimate of the likelihood that such a thing might happen naturally. Certainly the incidence of so many deaths occurring during Lucia's shifts does not seem likely, but it is not enough to simply say so. One has to do the mathematical calculation in order to decide whether it is so unlikely as to virtually rule out its happening by coincidence, or whether it is simply at the margin of reasonable probability—a combination of events that, although rare, can still be expected to occur now and then in every country.

Indeed, if we imagine making a list of all the thousands of nurses in a country, together with the number of deaths at which each one was present, then there will be at least one unfortunate nurse who will find herself at the extreme end of that list, with an unusually large number of deaths compared to the others. But surely we are not going to arrest her! The purpose of the calculation is to quantify the difference between a nurse's being in this position—at the extreme end of a natural curve—and her being far beyond the edge of the natural curve, a likely murderer.

To make sense of the computation, however, requires an experienced and careful statistician, which Henk Elffers was not. What Elffers did was to apply a standard statistical test called "Fisher's exact test" on the figures presented in table 7.1. This test yields the *p-value*, a number between 0 and 1 that tells you the probability that a set of numbers has of being an absolutely random distribution with no outside influence. For instance, a p-value of 0.05 or greater means the situation lies within the range of events that occur more than 95 percent of the time. If the p-value is less than 0.05, this means the event one is considering lies in the marginal set of events that occur less than

5 percent of the time. A p-value of 0.01 means the combination of events being considered occurs about 1 time in 100. As we already saw in the p-value discussion in the Berkeley admissions case (see chapter 6), a p-value of under 0.001, less than 1 in 1,000, is accepted as small enough to raise suspicions about the naturalness of the event under consideration, but not sufficient to conclude that there is definitely something wrong.

When Elffers performed Fisher's test on Lucia's table for the Juliana Children's Hospital, it yielded a tiny p-value of 0.000000110572, or less than 1 in 9,000,000. This p-value of 0.000000110572 would indicate that the table of shifts and deaths concerning Lucia's work at Juliana Children's Hospital corresponds to a combination of events that would occur only once in about 9,000,000 cases of a nurse working at a hospital over a nine-month period. Since there were 27 nurses at the hospital, Elffers multiplied this p-value by 27 to obtain the chance that such a combination could occur within the hospital, obtaining a result of 0.0000029854, about 1 in 350,000. Definitely an event rare enough to provoke worry in a country that contains only about 250,000 nurses in total.

Elffers' next step was to apply the same test on the data for the two wards at the Red Cross Hospital where Lucia had also worked. There, he used the following tables. For the first table (table 7.2a), which showed that Lucia had worked roughly a quarter of all shifts and been present at roughly a quarter of all incidents, he calculated a p-value of 0.07155922, so about 1 in 14, making this table appear well within the bounds of normalcy.

The second table (table 7.2b) shows that during the nine-month period under investigation, 5 patients died, and 366 nursing shifts were worked. During that period Lucia worked just a single shift, and during that one shift a patient died. No special probability test is necessary to see that under the assumption that shifts and deaths were randomly distributed, the probability of this happening is about 5/366, 0.0136, or about 1 in 73.

When the three p-values of 0.0000029854, 0.071559, and 0.0136 are multiplied together, the product is close to 1 in 342 million. In a letter to the *Guardian* of April 10, 2010, Henk Elffers denies having multiplied the values together in this manner. However, a memo authored by him and dated May 29, 2002, gives the above calculation in detail. Furthermore, the argument Elffers gave in court indicated that this multiplication should be carried out; indeed, it was performed by the prosecution and quoted to the

TABLE 7.2a

Red Cross Hospital Shifts / Ward 1	Number of Shifts without Incident	Number of Shifts with Incident	Total
Without Lucia	272	9	281
With Lucia	53	5	58
Total	325	14	339

TABLE 7.2b

Red Cross Hospital Shifts / Ward 2	Number of Shifts without Incident	Number of Shifts with Incident	Total
Without Lucia	361	4	365
With Lucia	0	1	1
Total	361	5	366

judge, and it was cited in every European newspaper covering the case. The figure was interpreted to mean that there was *a single chance in 342 million* that such a distribution of deaths and shifts could occur naturally. Since 342 million is many times the number of nurses on the planet, that meant that this distribution was not likely ever to occur naturally at all. But it *had* occurred—therefore, the conclusion that Elffers drew was that it simply could not have been natural.

Elffers explained to the court that conceivably other, non-malicious factors might have caused the strangely skewed results. He suggested five possibilities: (1) perhaps Lucia was a particularly incapable nurse; (2) perhaps she was systematically assigned to the patients with the worst health; (3) perhaps she was given or chose special shifts with respect to those of her colleagues, for example the night shifts, during which the majority of patients die; (4) perhaps there was some other person who was also present at every single one of the suspicious incidents; and finally, (5) perhaps someone was trying to frame Lucia.

But Lucia rejected all of these suggestions in her testimony. Without meaning to incriminate herself, but simply telling the truth, she testified that she was a good nurse and that difficult shifts and patients were shared equally among all the nurses. No one else had been present at all the deaths. She did not think anyone was trying to frame her. She believed that what had happened was simply a matter of chance, in spite of the numbers. But her belief was contested by expert witness Elffers. Turning to the judge, he stated: "Your honor, it was not chance. The rest is up to you."

During the trial, Lucia de Berk was treated by the press like a monster. Her denial of all wrongdoing and her refusal to confess made her seem even guiltier, and she became a focal point of loathing for an entire country. The revelation that she had worked as a call girl in Canada for a couple of years was just more grist for the media mill. She was accused of all kinds of things, from arson, to stealing books from the hospital li-

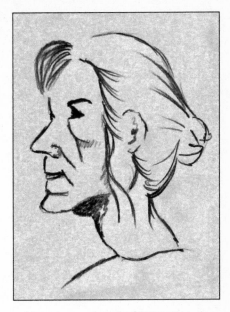

Newspaper sketch of Lucia in court

brary, to selecting mainly Arab children as her victims in the belief that the Islamic parents would object to autopsies. Even the drawings of court proceedings published in the newspapers showed a witch-like figure with no resemblance to the real woman.

On March 24, 2003, the court in The Hague sentenced Lucia to life imprisonment for four murders and three attempted murders. Her defense team had succeeded in having the original list of accusations reduced from seventeen to seven—they had actually shown that she had not even been present for several of them. But it wasn't good enough.

LUCIA APPEALED the verdict, and the case was retried at the High Court of the Hague in June 2004. The prosecution amassed a string of new evidence on the various cases, providing what was considered to be incontrovertible proof of the murders of Amber and Achmad. They also presented, with great fanfare, the testimony of a fellow detainee of Lucia's who asserted that one day during exercises in the prison yard, he had heard her say that she had "released these 13 people from their suffering." Interrogated in the courtroom, however, the man admitted that he had invented the entire story.

On June 18, 2004, Lucia was now found guilty of seven murders and three attempted murders—four new murders had been attributed to her on appeal, while only three of the four original murders for which she had been convicted in the first degree were among the new seven. Indeed, Lucia's defense had succeeded in finding information proving that in the original list there was yet another case at which Lucia had not actually even been at the hospital, but was absent on holiday leave. After this discovery, that particular death quietly disappeared from the list; no one asked any longer whether it had been natural or unnatural. But neither did anyone recalculate the p-values of the updated tables.

The set of deaths that were qualified as murders kept changing according to whether or not Lucia was known to have been present. This already should have been a red flag: it wasn't merely a matter of *who* had murdered the patients at this point, but whether they had been murdered at all. In essence, if Lucia had not been on shift, deaths were considered natural, but if she had been present, they were murders—the status of each death seemed to hinge on whether or not Lucia had been present. The whole affair was turning into quicksand.

In its judgment the court explained that it was considering only the cases of Amber and Achmad as murders that absolutely could be proven by medical evidence. But it relied on the "chain-link" hypothesis: if two cases were murders, then the others must be as well. As for the statistics, the court stated firmly that despite appearances, "a statistical probability calculation played no part in the conviction." The evidence they considered, they insisted, was entirely medical. And they had any number of medical doctors on the stand, explaining exactly why they believed that the deaths were unnatural—explanations all the more necessary for those doctors who had originally certified the deaths as natural. They changed their minds, and in the end their statements convinced the court.

On appeal Lucia was given one of the harshest possible sentences in the Netherlands: life imprisonment together with coerced psychiatric treatment, in spite of the fact that the psychiatrist who had followed her for six months during her imprisonment found no evidence of mental illness at all.

Three days after the judgment, a new report on the contents of the swabs containing Amber's diluted blood arrived from a laboratory in Strasbourg with state-of-the-art techniques for detecting proportions of digoxin. Since the trial was over, the report was shelved, but it was taken out and submitted as new evidence when, Lucia having appealed again, the case came before the supreme court of the Netherlands. The document was not examined by the court, however, which ruled on March 14, 2006, that life imprisonment and coerced psychiatric treatment could not be combined. The case went back to the Amsterdam court of appeals to be judged again on the basis of the same facts as before. The court upheld the guilty verdict for the seven murders and three attempts.

After hearing this verdict destroy her last chance at freedom, Lucia suffered a stroke. She lay paralyzed in her cell for ten hours before finally being transported to the penitentiary hospital; having never seen her ill previously, the prison guards and nurses believed she was putting on a hysterical act. After this lengthy delay in treatment, Lucia lost the power of speech and all motion on the right side of her body. There was nowhere left for her case to go; she had no more hope.

LUCIA DE BERK would likely have spent the rest of her life in prison if it had not been for a group of siblings who became deeply involved in her case. Geriatrician Metta Derksen de Noo was the sister-in-law of Arda Derksen, the chief pediatrician at Juliana Children's Hospital who had made the first list of suspicious incidents at which Lucia had been present for the hospital's director, the list that he took directly to the prosecutor. Arda had subsequently overseen the internal hospital investigation that preceded the report of murders and attempted murders to the police. Like Paul Smits, she also tried to work out the statistical probability of the deaths during Lucia's shifts having occurred by pure chance; she was aided in this endeavor by a theoretical computer scientist from Amsterdam, who was also a member of her family. Later, Arda was also helpful to the prosecutor. She had been suspicious of Lucia even before Amber's death, due to the "unexpected" resuscitation of one of her own patients. Once the hospital was full of gossip about Lucia, Arda's suspicions had become stronger than ever.

Yet during Lucia's trial, Arda suffered from some health problems that made her unable to testify. Her behavior increased the suspicions of her husband's sister, Metta de Noo, that something might be wrong with the case to which Arda was devoting so much attention.

In the winter of 2004, after Lucia's first conviction but before her appeal, Metta began to investigate all of the medical records of the case—all the records concerning the supposed deaths of patients at Lucia's hands and the transcripts from her trial. She was shocked to realize that many of the medical diagnoses she was seeing appeared unconvincing, even insincere; in fact, she could not point to one solid piece of evidence that indicated murder at all. Disturbed by Arda's problems and upset by these circumstances, Metta intensified her investigations, wrote innumerable letters, and repeatedly visited lawyers and doctors about the case—with no results. In November 2005, she began consulting with Lucia's lawyer, and later she became personally acquainted with Lucia in prison, created a website devoted to the case, founded the Committee for Lucia de B. in her support, and got her own husband, her mother, and two of her brothers involved in Lucia's defense.

Although Metta found little support during the first year or two of her efforts, she gained a powerful ally when she finally persuaded her brother Ton to champion the cause. Ton Derksen, a professor at the Faculty of Philosophy of the Radboud University in Nijmegen, was able to get the ear of authorities such as the public prosecutor, who had remained deaf to Metta's demands. The detailed investigation that Metta had started into all of the medical, personal, and statistical evidence culminated in a remarkable book authored by her brother Ton, *Lucia de B.: Reconstruction of a Judicial Error*, as well as more recent books by both Metta and Lucia herself.

Although none of these books have yet been translated into English, a translation of part of Ton's book can be found on the Internet; it contains a full-fledged crime investigation performed in extreme detail. Much of the work it contains calls the skill of Lucia's lawyers into question as, even while working for her in good faith, they missed several important pieces of information, and like the prosecution, they relied on mathematical expert witnesses who were not professional statisticians and thus not fully equipped to refute the testimony of Elffers and his colleagues.

In his book, Derksen examines the case of Amber, the main locomotive case used to convict Lucia, and shows that the proof of her murder reconstructed by the court included two major factual errors. The first concerned the time at which the court concluded that Lucia had administered the drug to the baby. On September 3, Amber had appeared unwell to the nurse, as she was needing increasing levels of oxygen to help her breathe. At 11:00 that night, Lucia had connected Amber by finger cuff to a monitor to help follow her oxygen saturation levels, and at a time that she recalled as being "about 1:00 a.m.," she called for a pediatrician to come and examine the sick child. So Amber was disconnected from her monitor and wheeled into a special room. Two doctors examined Amber at a time noted in the register as 1:00 a.m. According to their testimony, the examination lasted "about 20 minutes," after which they concluded that the child was not seriously ill and sent her back to her room, where she was reconnected once again to the monitor. The graph from the monitor does indeed indicate an interruption at around 1:00 a.m., but it is a short one, corresponding to a period of five to ten minutes rather than the twenty to twenty-five minutes it would have taken for Amber to be examined, counting the transport from her room to the examination room and back.

Amber's crisis began at 2:46 a.m. with two nurses in the room. The re-suscitation team was summoned and did all they could, but she died a short time later.

A person dies sixty to ninety minutes after an overdose of digoxin, so the court considered the printout from Amber's monitor from 1:15 a.m. to 2:45 a.m., and they found a lengthy interruption of monitor activity lasting from about 1:20 a.m. to 1:48 a.m. This, the court judged, must correspond to the murder time: Lucia must have turned off the monitor and introduced the poison into the intravenous tube that had been placed on the baby at her examination. Her conviction for Amber's murder was based on this conclusion together with the presence of traces of digoxin found in the child's body.

But Derksen points out that the medical examination could not have taken place at 1:00. The interruption was too short: there simply would not have been enough time. He attributes the brief interruption to the fact that the nurse was probably cleaning or changing Amber, who was suffering from acute digestive difficulties. A careful inspection of the monitor graph indi-cates that only the interruption from 1:20 to 1:48 is long enough to corre-spond to the doctor's examination: Derksen concludes that this examination must have begun around 1:20 rather than the imprecise 1:00 noted down in the register. There was nothing unusual in this; indeed, Derksen also observed that every one of the times noted in the register was either on the hour or on the half hour; it was not usual to write down the exact time to the minute. The doctor's estimation of the hour of their examination was simply off; it must have taken place at 1:20 rather than 1:00 a.m.

This, together with the known delay in effect of digoxin, would imply that Lucia must have poisoned the baby *during* the medical examination. She could not have done it later, since the monitor graph shows continuous activity from 1:48 until the end of Amber's life. Nor could she have done it before, as death would then have occurred earlier than it did. As Derksen pointed out, it is virtually impossible to pinpoint a time when the murder could have actually taken place.

The second major error in the court's proof of Amber's murder was the medical evidence of digoxin poisoning. To show that no such poisoning had occurred, Derksen used the report from the Strasbourg laboratory that had been rejected by the supreme court.

The difficulty with measuring digoxin concentrations is the presence of DLIS, Digoxin Like Immunoreactive Substances, which can occur naturally in the body, in particular in the bodies of infants. The two medical tests whose results were used by the court were not able to distinguish between digoxin and DLIS; they detected concentrations of 22 and 25 milligrams per liter of digoxin in Amber's blood, contrasting with an expected level of 1–2 milligrams per liter. However, digoxin concentration increases after death—almost any concentration will increase with evaporation of liquid—and it is considered that a finding of 1–7 milligrams per liter indicates an absence of any poisoning.

This very concentration, 7 milligrams per liter, was exactly the result found by the Strasbourg laboratory, which used the only known method capable of distinguishing between digoxin and DLIS. This information had been ignored by the Dutch supreme court, along with several other medical findings from Amber's autopsy, such as non-contraction of the heart, which indicated that she had not died from digoxin poisoning at all.

Amber's short life was tragic enough. But she was not murdered.

WITH THE main locomotive case destroyed, the case for the other murders was greatly weakened. Yet a major question remained: why was Lucia present at such an unrealistically large number of deaths? The result of Elffers' calculation shows that it is virtually unthinkable that such an event could happen by chance, as Lucia claimed. Smits' calculation, confirmed by Elffers' p-value, was the basis of the original suspicions against her. How could it be explained?

Fortunately for Lucia, a battery of professional statisticians—including Professors Richard Gill and Peter Grunwald of Leiden University—joined Metta de Noo and Ton Derksen in their battle for the rectification of injustice. On closer examination, these professors concluded that the statistical analysis methods used by Henk Elffers, although convincing at first sight, were in fact inaccurate and not correctly applied to the case. One of the most glaring errors is the multiplication of the p-values (which Elffers denies having done, but the result of which reached both the court and the newspapers).

Although p-values can be multiplied under certain circumstances, such as the total independence of the events whose p-values are being measured, they cannot and should not be multiplied as a general rule. A simple example illustrates why multiplying p-values can give a drastically wrong result.

Imagine a situation similar to Lucia's, but simplified. Suppose that nurse N, while not a bad nurse, is a little more careless or perhaps less trained than most other nurses and makes occasional mistakes in her treatment of patients. Every nurse probably makes such mistakes from time to time, and the large majority of these are not harmful. There are about 250,000 nurses in the Netherlands; let's suppose that without doing anything drastically bad, nurse N sits squarely in the middle of the group of 1,000 nurses who make the most errors. If you make a table containing all the errors made by N in the hospital where she works, you will find a p-value of 1/250, meaning that N belongs to the group of 1,000 nurses who make the most errors among the 250,000. Since this p-value is greater than 1/1,000, it is above the level accepted as giving rise to legitimate suspicion.

Now suppose that nurse N works in two hospitals. Given the kind of nurse she is, if two separate p-values are calculated in the two hospitals, both will probably be close to 1/250, and both will point to the same fact: that nurse N has some problems with accuracy. But what happens if you multiply them? Suddenly you get a p-value of 1/62,500, meaning that N must be one of the very worst nurses in the country, with a p-value that may legitimately arouse suspicion as to her acts.

Yet clearly the fact of working in two different hospitals should not change anything in the calculation of nurse N's p-value, since obviously she

is the same nurse in both hospitals. The reason for the error is that her relatively poor performances in the two hospitals do not constitute independent
events, since both stem from the same underlying cause, which may be inat-
tentiveness or poor training. This is why it is wrong to multiply the p-values:
in this case, that operation does not reflect reality.

Lucia denied that she was in any way the cause of any of the lethal or
near-lethal events at which she had been present. She attributed the number
of deaths in her presence to absolute random chance, and if she was right,
then multiplying the p-values would not have been an error. So one may ask
whether she, like nurse N, may have been a nurse whose performance was
within the bounds of normality but still under average—in which case the
p-values should not have been multiplied, and thus the statistic most indicative of her guilt would be invalid—or whether she was in fact right, and
the defect in Henk Elffers' calculation was something different.

Over the course of his investigation, Ton Derksen made an interesting
discovery: there was, indeed, another serious flaw in the calculation made
in court. It wasn't in multiplying the p-values, it wasn't even in any of the
mathematical operations; it was something more fundamental: the compi-
lation of the actual tables. Elffers appears never to have double-checked
them, but Derksen dug into the actual making of these tables.

The hospital claimed to have first made a list of suspicious incidents and
then proceeded to check whether Lucia had been present—only to find that
she had been there for all of them! This sounds like a simple procedure with
a damning result, but it is in fact misleading. The ambiguity lies in how the
incidents were classed as "suspicious." Since no one had suspected anything
when each incident occurred, they were termed "suspicious" only in hindsight. But a list of some nine or ten incidents at which Lucia had been present was circulated on the day following Amber's death, so—as Derksen
asked—how is it possible to believe that the list of "suspicious" incidents
was made without any knowledge of her presence?

The real question is whether there were any suspicious incidents that
occurred when Lucia was not present, incidents that, like those on the list,
were not considered suspicious at the time but had similar features. If there
were any such incidents that had not been included in the original list, this
could change the table and have a considerable effect on the p-value.

It was not an easy question to answer, since none of the events had been
found suspicious at the time, but as Derksen searched through the hospital

TABLE 7.3

Juliana Children's Hospital Shifts	Number of Shifts without Incident	Number of Shifts with Incident	Total
Without Lucia	887	0	887
With Lucia	134	8	142
Total	1021	8	1,029

TABLE 7.4

Juliana Children's Hospital Shifts	Number of Shifts without Incident	Number of Shifts with Incident	Total
Without Lucia	883	4	887
With Lucia	136	6	142
Total	1,019	10	1,029

records, he did find a few striking things. Consider, for example, the case of the child Kemal, who underwent three resuscitations and a coma from an overdose of chloral hydrate. Two of the resuscitations found their way onto Smits' famous list, whereas the third was left out. Why was this? Derksen could find no differences in the circumstances surrounding the three resuscitations, except for one: Lucia was not present at the third one, and that one was absent from the list. As for Kemal's coma, it was closely related in type and in cause to Achmad's coma, which was also the result of an overdose of chloral hydrate. But Kemal's coma was not classified as suspicious. That is, Lucia had not been there. Derksen went on to discover two further incidents that were exactly as suspicious as those on the list but had taken place in Lucia's absence.

Recall the table compiled by the investigators at Juliana Children's Hospital and presented at her first trial (table 7.3 above). We can correct this data by removing the two incidents that were discovered to have occurred outside her shifts and subsequently removed from consideration at her trial, and then adding the two incidents above concerning Kemal as well as two others discovered by Derksen. The new table looks like table 7.4.

Applying the same method as before (Fisher's test) to table 7.4 yields a p-value of about 1 in 1,230. This is a far cry from the p-value of 1 in 9,000,000 corresponding to the first table! A figure of 1 in 1,230 is, as we have said, sufficiently small to arouse some suspicion, but here, given the number of nurses in the Netherlands, it is really not so unusual. It simply means that out of some 250,000 nurses, one could expect a couple of hundred to be involved in a set of circumstances similar to those of Lucia. And indeed other nurses did come forward with their own stories: a letter to the newspaper written by a nurse in support of Lucia's situation mentioned that during her student years she herself was present at thirty deaths, whereas a fellow student of hers was not present at any.

ON THE basis of his discoveries, Ton Derksen submitted a request to a committee, the Buruma Committee, to decide whether Lucia's case should be referred to a special body called the Committee for the Evaluation of Closed Criminal Cases. No defendant is allowed to submit such a request; it must emanate from someone external to the case.

The Buruma Committee recommended further investigation, and by October 2006 a panel of three was appointed to conduct research into the situation from scratch. Above all, they were asked to focus on certain aspects of the case—namely, to determine:

- Whether the statistical evidence was supported and complete.
- Whether the question of digoxin poisoning had been fully resolved.
- Whether the only unexplained deaths at the hospital were truly all under Lucia's supervision, or whether other deaths had been put aside on the basis that they did not involve her.

The Grimbergen Report, named for one of its authors, was released in October 2007 after ten months of work. It contains a sort of apology for the length of time it took to put it together; the three researchers had wanted to do the best possible job, but at the same time they were well aware that Lucia was spending those months in jail in a state of increasingly poor health.

In the report, they note some of the major errors in the initial investigation: the fact that after Amber's death, Lucia was considered a suspect—the only suspect—almost immediately, and the fact that this led to a conscious choice on the part of the investigators to focus the investigation on the ward where she worked during the period she worked there. In fact, Lucia had been suspected of wrongdoing even before the death of Amber. She was aware of this suspicion and had wondered whether perhaps Amber had been entrusted to her care as some kind of test. This focus on Lucia's guilt led the police to miss or not take account of simple observations of major importance. For example, in the ward of Juliana Children's Hospital where Lucia worked, during the period she worked there, there were six patient deaths, whereas in the period of equal length before she arrived, there had been seven! Accusing Lucia was tantamount to assuming that the number of deaths had actually dropped once an active serial killer arrived on the scene. It was such a simple remark, yet it had had no influence at her trial.*

The report also investigated the additional resuscitation incidents at the hospital that Ton Derksen had unearthed, as well as the Strasbourg report about the measurement of digoxin in Amber's blood. The researchers concluded that not only had the prosecution seriously erred in concluding that the child had been poisoned, but that the defense, too, had erred in accepting that conclusion. The report stated that there seemed to be no indication at all that Amber had died from digoxin poisoning. One of the world experts on digoxin, Professor Gideon Koren from the University of Toronto, wrote a letter after having examined the evidence in detail, and his conclusion speaks for the report: "I'm of the opinion that every attempt to interpret the post mortem level as proof of poisoning (inadvertently or on purpose) is incorrect and, in all honesty, quite shocking. The idea that a professional in health care could be imprisoned because of such an incorrect interpretation would be absolutely unacceptable."

The Grimbergen Report recommended that public prosecutors initiate a request for judicial review of Lucia's case. A petition was immediately handed to the Dutch minister of justice asking for Lucia's detention to be temporarily suspended during the revision of the case, but this request was refused. The petition appeared the following day as a full-page newspaper advertisement.

---

*The report also showed that Arda Derksen, chief pediatrician at Juliana Children's Hospital, was the contact person at the hospital for the investigation by the public prosecutor. The medical records were all confidential, but she communicated their contents to the prosecution via summaries.

Lucia de Berk, a free woman

On January 5, 2008, a "Light for Lucia" torch procession was held at the prison where she was incarcerated, and a month later a new play called *Lucy, a Monster Trial* premiered in Amsterdam. The supertanker of media coverage and public opinion was slowly, hulkingly, beginning to turn around.

Lucia appealed the decision to refuse her liberty, and on April 2, 2008, the minister of justice granted her a temporary suspension of sentence for three months, on the basis that the proof of Amber's murder had melted into thin air. Although still partially paralyzed by her stroke, Lucia was able to walk out of Nieuwersluis prison. It was her first day of freedom in more than six years.

Events followed with the excruciating slowness typical of the workings of justice, but each successive step led closer to the final truth. In June, the Dutch supreme court was officially requested to reopen the case; in October, it accepted and began by ordering a set of entirely fresh medical investigations, not for Amber but for other cases—in particular, Achmad, the other "locomotive," and Achraf, the child whose sudden need for resuscitation had been so unexpected. Fourteen months later, in December 2009, the court accepted the testimony of the new medical witnesses to the effect that the deaths now appeared to be entirely natural. The trial was adjourned a final time, to March 17, 2010, the day that Lucia herself was interrogated for the last time. At the close of the trial, the public prosecution itself, in what must surely be a historic gesture, requested the court to acquit her.

On April 14, 2010, the court delivered its final verdict of not guilty. In doing so, it reversed a miscarriage of justice that devastated lives and reputations from that of the justice system of the Netherlands, in the eyes of the people and the international community, to that of a humble woman who had only ever wanted to be a nurse, and who had been a good and cherished one until her world came crashing down around her.

MATH ERROR NUMBER 8 »
UNDERESTIMATION

An ancient Indian legend tells the story of a *dravida vellalar* who invented the game of chess. The sultan was so delighted by the game that he offered the inventor any gift he might wish. The *vellalar* asked to receive grains of wheat, counted using the 64 squares of the chess board, by placing one grain on the first square, two grains on the second square, four grains on the third square, eight grains on the fourth square, and so on, doubling the number of grains for each square. The sultan, slightly offended at the measly nature of the gift, accepted, and ordered his treasurer to count and hand over the wheat.

Several days passed and no wheat was forthcoming, so the sultan called for his treasurer and asked what the problem was. The treasurer told him that the amount of wheat required would form a heap incomparably greater than the highest mountain on Earth.

Vexed, the sultan called for the inventor and told him that his wheat was ready, and he should go and count every single grain before taking it away.

THE REASON FOR the sultan's error in this story is a widespread lack of intuition concerning the speed of exponential growth. The rational, commonsense part of our brains is accustomed to observing a small portion of something and then extending that image to form a mental picture of the whole. In the case of the chessboard, the sultan probably thought something like: 1 grain, 2 grains, 4 grains, 8 grains, 16 grains, 32 grains, 64 grains, 128 grains, 256 grains . . . Why, this poor beggar will barely get enough for one dinner!

Our daily habits of working with relatively small amounts, small distances, and small sizes create a sort of mental block against gigantic numbers. Our lack of experience prevents us from having the kind of realistic, familiar understanding of their meaning that we have for the numbers that measure quantities we frequently encounter. We build our mental pictures around what we are used to, and this can lead to great surprises when dealing with things or quantities that are extremely large.

Here is another example of this sort of error: a problem whose answer astonishes even professional mathematicians.

Assume that the world is a perfect sphere and wrap a wire tightly around the equator. Now take a second wire that is exactly one meter longer than the first one, and wrap that around the equator as well. Because this wire is longer, it will be slightly loose, a bit up off the ground all the way around. But how high will it actually be? Can you slip a razor blade underneath?

Before calculating anything or looking at the answer below, make a picture of this situation in your head. You're standing on the ground, and the first wire is tightly running along the equator at your feet. The longer wire, with a meter added in, runs along the equator too, but just a little bit off the ground, since it's just a bit longer. Mentally look at your feet. See the wire near your toes? How loose is it now? How high off the ground? How much difference did one little meter make compared to the whole equator of the earth?

Here's the answer: the second wire is almost 16 centimeters off the ground, all the way around the earth. Not only razor blades, but a row of rabbits, can fit underneath it from Ecuador to Malaysia and back around the other way.*

---

*How do you prove this? The simple circumference formula for a circle: $C = 2\pi r$. Here C is the circumference of the earth, which is about 40,000 km, and $\pi = 3.14159$. So assuming the earth is a perfect sphere, the radius r = 6,366.197722 km. The length of our original wire is 40,000 km, and we add one meter to that length. A meter is one-thousandth of a kilometer, so the new length of the wire is 40,000.001 km, and we divide this by 2π to obtain the radius of the new circle formed by the lengthened wire. As expected, the new radius is very close to the old one; the calculation gives the answer 6,366.197881 km. The difference between the earth's radius and the new one corresponding to the longer wire is just the difference between the two—the tiny-looking number, 0.000159 km. This is equal to 0.159 meters, or almost 16 centimeters.

This answer is incredible to our minds; it seems to us that one meter is so tiny compared to the circumference of the earth that it should make essentially no difference at all to the height of the wire. In fact, we are seriously underestimating the true "looseness" that occurs. The reason for this is that the circumference of the world is too large for us to intuit and to compare correctly with a small dimension like a meter. We think of 16 centimeters as being "much too big," although in fact it is minuscule compared to the earth's radius.

Getting back to the chessboard story above, let's take a look at the numbers. After the sultan has finished counting off the grains corresponding to half the chessboard, 32 squares, he is already up to 4,294,967,295 grains of wheat, weighing maybe 100,000 kilograms. That much wheat can fill up enough boxes to stock the shelves of a thousand supermarkets, and it is certainly already well over what the sultan had thought he was bargaining for. After 64 squares, there would be 18,446,744,073,709,551,616 grains, nearly 500 billion metric tons, which is comparable to the mass of a large mountain.

The important point here is to recognize an *exponential growth pattern,* or any growth pattern that doubles (or triples or whatever) at regular intervals. Such a growth pattern starts out relatively slowly, and then accelerates with frightening speed. For instance, be wary of participating in e-mail chain letters in which you are incited to send ten copies to your friends on pain of some curse falling on your head. If everyone who received the letter followed the instruction, the letter would end up spamming more than the population of the world in ten reiterations, clogging the world's bandwidth with unspeakable amounts of virtual clutter.

The case we are about to describe is a story of people who got fooled because they did not realize the implications of the incredible rapidity of exponential growth.

The Case of Charles Ponzi:
American Dream, American Scheme

It was in the cold month of November 1903 that twenty-one-year-old Carlo Ponzi arrived in Boston, Massachusetts. He had sailed from Italy, which was nothing unusual, for Italy at the turn of the century was one of the most massive sources of emigration to the New World, whether to Canada, the United States, or South America. The only thing that differentiated Ponzi from the majority of the other immigrants was the fact that he had a university degree, from the University of Rome, La Sapienza. However, not unlike many university students today, Carlo had discovered soon after graduating that a degree earned by dint of assiduous frequentation of bars and gambling halls did not easily lead to a situation of gainful employment. Not that a situation of mere gainful employment held much charm for Carlo. He was a man of ambition, and a mere salary did not attract him. If he left for America, it was because he meant to realize the American dream.

He brought a little money with him but was fond of telling the story of how he gambled it all away on board, so that on disembarking, he had only $2.50 in his pocket—"but $1,000,000 in hopes!" One of his first acts upon settling in the United States was to change his name from Carlo to Charles; he then set about looking for a job. But the reality of his situation soon hit: penniless and speaking broken English, Ponzi could find no other work than waiting on tables and washing dishes. A first, cheap attempt to earn extra money by shortchanging customers got him fired.

In the year 1907, the United States went through a sudden economic crash known as the "panic of 1907" (the same crisis during which Wall Street financier Hetty Green, whose story is told in the following chapter, loaned more than a million dollars to the city of New York). The New York Stock Exchange fell 50 percent from its peak the year before, and the images that we associate with the 1929 depression—financial institutions falling into bankruptcy like dominos, crowds of people assailing the banks in a desperate bid to get their money out—were familiar sights in 1907.

The crisis of 1907 was finally stabilized thanks to the intervention of a single figure: J. P. Morgan. Considered a hero at the time, Morgan later came under heavy criticism for his excessive role in the financial life of the United States. A committee was formed to investigate his "money trust," and Morgan was called in for questioning. The resulting dialogue went down in the annals of financial history:

> COMMITTEE QUESTIONER: Is not commercial credit based primarily on money or property?
>
> J. P. MORGAN: No, sir. The first thing is character.
>
> COMMITTEE QUESTIONER: Before money or property?
>
> J. P. MORGAN: Before money or anything else. Money cannot buy it . . . a man I do not trust could not get money from me on all the bonds in Christendom.

Charles Ponzi might have been listening in, so well did he learn the lesson imparted by Morgan.

THE FINANCIAL situation in the United States in 1907 made it impossible for Ponzi to find work, and he decided to venture north to Canada. He landed in Montreal, where he contacted an Italian immigrant, Luigi Zarossi, who ran a successful cigar business on the rue Saint-Jacques. Introducing himself as Carlo Bianchi from a nonexistent (but wealthy) Italian family, Ponzi was befriended by Zarossi, who offered him a low-level job as assistant bank teller at Banca Zarossi, the bank he had founded to service the growing ranks of Italian immigrants who were arriving in the city.

Thanks to some impressive real estate deals, Banca Zarossi was able to offer its clients 6 percent interest, more than double the going rate at the time. Ponzi developed a close friendship with Zarossi and his family, and rose to the position of bank manager. It was there that he learned his first major lesson in finance. The bank was growing so rapidly that Zarossi's real estate holdings were not sufficient to pay out the 6 percent interest rate, and Ponzi discovered that Zarossi was using the money from new clients to pay it out instead. This meant that in case of a rush on the bank, Zarossi would be unable to refund money to his clients. When rumors about the situation began to spread and it became clear that Banca Zarossi was on the point of collapse, Zarossi stole all the remaining money and fled to Mexico, leaving his wife and children behind.

Ponzi moved in with Zarossi's abandoned family and made a sincere attempt to help them, but between their complete lack of money and the anger of the cheated clients, the situation was catastrophic. He went for help to the offices of a business called Canadian Warehousing, which had been one of Banca Zarossi's clients, but found no one there. Seeing a checkbook behind the desk, Ponzi seized the opportunity to write himself a check for the authentic-sounding if large sum of $423.58 (about $10,000 in today's dollars). He signed the check with the name of one of the directors of the company and cashed it.

Unfortunately, however, although Ponzi had not been charged with any crime in the collapse of Banca Zarossi, his doubtful position and penniless situation were well-known to Ponzi's neighbors, who alerted police when they saw him suddenly spending large amounts of money in the days that followed. He didn't try to hide it: as the police officers walked toward him, Ponzi famously held out his wrists and simply said, "I'm guilty." He was sentenced to a three-year prison term at Montreal's St.-Vincent-de-Paul Penitentiary. Released after twenty months for good behavior, Ponzi decided to return to the United States and "start afresh."

IF "STARTING AFRESH" meant seeking more opportunities to get rich quick, then it can be said that Ponzi made great efforts. Ten days after his arrival in the United States on July 30, 1910, he was arrested again, this time for participating in a plan to smuggle illegal Italian immigrants over the border. During the two years that he now spent in prison in Atlanta—probably a welcome change from Montreal, particularly in the wintertime—Ponzi, like others before and after, learned a remarkable amount about the art of criminal technique. He became friends with Ignazio Saietta, a Sicilian mafia mobster jailed for counterfeiting money (after dozens of murders), and Charles Morse, a powerful New York businessman, whose efforts to corner the stock of the United Copper Company had been one of the major causes of the economic panic of 1907. Morse was used to a life of wealth and ease; he knew his way around money, and on top of that he succeeded in fooling a panel of army doctors into declaring that he was so ill that if he did not immediately receive a presidential pardon to allow him to go abroad for treatment, he would die. But Ponzi knew the truth: Morse had ingested a drink of soapsuds before the doctors examined him. Soap is toxic and will produce symptoms of poisoning, but the toxins are generally reasonably harmless and soon pass out of the system. Morse left prison and took himself off to a German spa for "treatment." He left behind him a friend with a head full of new and interesting ideas.

Once out of prison, Ponzi moved slowly back north, following the jobs that he could find. He eventually landed a clerking position in Boston, and that is where he met and courted a charming petite Italian-American woman from a modest background, Rose Gnecco. Ponzi was now nearly thirty-five, and Rose was impressed by his sophistication and experience. Not that he told her the truth about his past—that was not one of his most noticeable habits—but Ponzi's mother, having learned of her son's relationship and feeling that the young girl ought to be warned, took it upon herself to write Rose a letter describing the main points of her son's hitherto unimpressive career. Rose did not care, or if she did, she did not change her plans, and the couple married in February 1918.

After his marriage, Ponzi's first job was to run his father-in-law's grocery shop, but this did not last long, although it was no fault of his: the business had already been on the brink of failure. Anyway, groceries were not his style; Ponzi wanted to make money fast, and he decided to start his own business. After some thought, he came up with the idea of an international trade journal, with profits to be made through advertising. Alas, his bank, the Hanover Trust Company, to which he applied for a two-thousand-dollar loan to cover his starting costs, rejected his application (with some prescience, it would seem), and the proposed journal, which had already been announced in various venues, came to nothing.

WHILE PONDERING his next move, Ponzi received a letter from a business correspondent in Spain who wanted to learn more about the nonexistent but much advertised journal. In order to facilitate a response, the Spanish businessman had included an international reply coupon (IRC) in his letter. The sight of it caused a stir in Ponzi's mind. This coupon was a way for the Spaniard to pay for Ponzi's return postage even though, being in Spain, he could not actually purchase an American stamp. It was an international system organized by treaty; the coupon could be bought for a few centavos in a Spanish post office and redeemed against a five-cent US postal stamp. The thing was, as Ponzi noticed, the Spanish purchase price was fixed, while the value of the Spanish currency against the dollar, like that of all other European countries, had fallen sharply in the aftermath of World War I. Ponzi saw an opportunity for what is known as *arbitrage,* taking advantage of a situation where the buying and selling prices of the same item are different for whatever reason.

Ponzi reasoned that the same opportunity would work in Italy, where he could obtain the help of members of his family. Converting $1.00 to Italian lire at the going rate of exchange would suffice for his family to purchase sixty-six IRCs* and send them over to Ponzi in large packages, where he would use them to buy five-cent stamps. In this way, he calculated, $1.00 would yield $3.30. Even subtracting commissions to family members and the cost of transporting the packages, it was still a tremendous profit!

Ponzi set up a company, the Old Colony Foreign Exchange Company, with offices on School Street in Boston. Its stated purpose was to organize the purchase and delivery of thousands of IRCs in Europe and exchange them for American stamps. The plan seemed flawless; the only trouble was capital. So he set out to find investors. And that is when Ponzi discovered his true talent, the one that would earn him millions.

IN ORDER to convince investors to put money in your company, you need three things: a convincing plan, a promising payoff, and above all—if one heeds the luminous words of J. P. Morgan quoted earlier—a personality that inspires trust. Charles Ponzi, perhaps one of the least trustworthy people the human race has ever produced, was able to inspire trust. His plan was infallible and legal. The promise he offered his first investors was unheard of: 50 percent interest in forty-five days, or 100 percent in ninety days. In effect, he was promising to double investments every ninety days. As for the trust, well, it seems that Ponzi could convince anyone of anything. After the first

---

*Roughly speaking, a dollar after the war was worth about 44 lire and an IRC could be bought for 1.5 lire. Of course the exchange rate was fluctuating, but 1919 marked an all-time low. In 1926 Mussolini curbed Italian inflation by pegging the dollar at just 19 lire.

few investors got their money back with the promised profits, Ponzi's doorbell never stopped ringing.

By April 1920, only three months into his scheme, Ponzi and his wife were living the kind of life they had only dreamed of. He bought expensive jewels for her, gold-topped canes for himself, and two cars. And his fortune kept growing. In May, he purchased a luxurious mansion in the bankers' part of the historic town of Lexington, near Boston, and a custom-built, chauffeur-driven limousine.

Charles Ponzi at the height of his success

And still the investors continued to fork over their money. What they didn't know, what nobody knew except Ponzi, is that he had stopped even thinking about buying IRCs. At the beginning, to be sure, he had had his family buying them in Italy and sending them to him in packets. But he found that redeeming huge numbers of IRCs in local post offices was difficult. The postal agents became suspicious of him and refused to give him the cash he demanded. Soon he was visited by government agents who informed him that any attempt to speculate on government-issued IRCs would be illegal. That was the end of that idea, although he never mentioned this incident to the investors.

By June, he had realized another of his dreams: vengeance. By buying a massive number of shares, he took control of the Hanover Trust Company, the very bank that had refused him a modest loan for his trade journal project the previous year. Never one to give up, Ponzi continued to pay interest to his investors with the huge flow of incoming money from new investors, all while telling himself that this reshuffling of money was merely a temporary ploy to tide himself over while he set up some serious investments. The new bank fell under the latter category; he also bought a meatpacking plant and real estate. But none of these ventures was anywhere near profitable enough to provide him with an interest rate of 100 percent every ninety days. So Ponzi kept on being Ponzi, handing money from new investors to pay the interest to the old ones. And that new money kept on rolling in.

By July, the scenes that met his eye as he pulled up at the office in his limousine each morning defied the imagination. He described them later in his autobiography:

> A huge line of investors, four abreast, stretched from the City Hall Annex, through City Hall Avenue and School Street, to the entrance of the Niles Building, up stairways, along the corridors . . . all the way to my office! . . .
>
> Hope and greed could be read in everybody's countenance. Guessed from the wads of money nervously clutched and waved by thousands of outstretched fists! Madness, money madness, the worst kind of madness, was reflected in everybody's eyes! . . .
>
> To the crowd there assembled, I was the realization of their dreams. . . . The "wizard" who could turn a pauper into a millionaire overnight!

If Ponzi managed to remain beloved even while the millions were coming in, it is because he always remained the modest and simple man he had been from the start: the true embodiment of a rags-to-riches fairy tale. At the peak of his triumph, in an interview with the *New York Times*, he described his story in the following terms.

> As I say, I landed in this country with $2.50 in cash and $1,000,000 in hopes, and those hopes never left me. I was always dreaming of the day I would get enough money on which I could make more money, because it is a cinch no man is going to make money unless he has got money to start on.
>
> I saved a bit of money from the odd jobs and had the time of my life for a couple of weeks. Then my cash was gone. So into the big town of New York I went to find a job. Up at one of the big hotels they needed some waiters, and they even furnished me with the tuxedo service coat. Yep, I've carried tons of food on the old waiter, and with the small salary and tips I made enough to live. I went from one waiting

job to another, worked in various hotels, small restaurants, and did my dish washing stunt from necessity at times. I got tired of New York and began to travel, getting jobs all along the way.

It was small jobs, and small jobs, up to the year 1917, when I headed for Boston. Once more, saw an advertisement in a Boston newspaper, answered it, and took a job with J. R. Poole, the merchandise broker. My salary was $25 a week.

And then I found my inspiration. She was Rose Gnecco, daughter of a wholesale fruit merchant of Boston, and the fairest and most wonderful woman in the world. All I have done is because of Rose. She is not only my right arm, but my heart as well. We were married in February 1918.

—*New York Times, July 29, 1920*

He comes across as a charming fellow, humorous, down-to-earth, friendly, and familiar with the struggles and difficulties of ordinary people. Of course, the reality differs significantly from the tale he told, but no one knew that. They wanted the American Dream, and Ponzi gave it to them. With his popularity and style, no one resented Ponzi's twelve-room mansion, his heated swimming pool, his limousine, his wife's diamonds. He truly made people feel that he deserved it all, and more—that they, too, could reach those heights of wealth if they just followed his Pied Piper call.

SUPPOSE NOW that you were an investor, wondering whether to invest in Ponzi's scheme. Back then our advice to you might have been: get in it as early as you can—wait just long enough to see if the very first wave of investors makes their promised profit. Get in early, and get out early. Take your earnings and run. In the twenty-first century, though, our advice would be more along the lines of: don't touch it with a ten-foot pole. Nowadays even successful investors in a Ponzi scheme are liable to lose more money than they earned, once the lawsuits start raining down. Why? Because any simplified model of Ponzi's scheme (more scam than scheme) shows that it is going to blow up, and pretty quickly. In 1920 no one should have been taken in; yet in 2010, people were still duped by such ploys, because the power of the dream is so strong. Only now, everyone suffers the consequences.

The first piece of advice we would give to someone considering an investment that promises astonishingly high rates of interest is to do a little calculation before making any decisions at all.

1) Money-Based Model

To make a simplified model of Ponzi's scheme, we will assume that every investor puts in $1,000, and that all of these investors ask for their interest and capital back at the end of the ninety-day period. In reality, many investors actually left their capital with Ponzi for a new round, so that he did not need as many new investors for each session; in this sense, Ponzi's scheme could last longer than our theoretical model. On the other hand, we also assume that Ponzi is not making any personal profit from the scheme but keeping all the invested money rolling among investors, whereas in reality, Ponzi was skimming off a portion of the incoming money for his personal use, which would cause his scheme to collapse sooner than our model. We consider that these two simplifying assumptions roughly cancel each other out.

The scheme is working as long as Ponzi has enough money at his disposal from new investors to pay back interest and capital punctually to all the earlier ones. The scheme collapses when he no longer has enough money to do that. The goal is to double everyone's money every ninety days.

We'll suppose that Ponzi started with 100 investors each putting in $1,000 for a total investment of $100,000. (In reality, it took Ponzi a few weeks to reach this point.) Ninety days later, he owes these investors $100,000 in interest as well as their capital back; he needs to find 200 new investors to put in $200,000 within the first ninety-day period, which he can pay out to the original investors at the end of the period. This accomplished, he now needs to double the $200,000 in order to be able to pay the 200 new investors back their money and interest at the end of the second ninety-day period. Thus, he must find 400 investors during the third ninety-day period to pay the 200 back, 800 in the fourth ninety-day period to pay the 400 back, and so on, with the number of new investors needed doubling every period. The moment he can't find enough new investors to enable him to pay back the previous ones, people will get scared; investors will start demanding their capital back, but new investors will not be forthcoming to provide the necessary funds, resulting in the collapse of the scheme.

TABLE 8.1 MONEY-BASED MODEL

Period	New Investors	Total Investors	Reimbursements
Start	100	100	
Period 1	200	300	$200,000
Period 2	400	700	$400,000
Period 3	800	1,500	$800,000
Period 4	1,600	3,100	$1,600,000
Period 5	3,200	6,300	$3,200,000
Period 6	6,400	12,700	$6,400,000
Period 7	12,800	25,500	$12,800,000
Period 8	25,600	51,100	$25,600,000
Period 9	51,200	102,300	$51,200,000
Period 10	102,400	204,700	$102,400,000
Period 11	204,800	409,500	$204,800,000
Period 12	409,600	819,100	$409,600,000
Period 13	819,200	1,638,300	$819,200,000
Period 14	1,638,400	3,276,700	$1,638,400,000
Period 15	3,276,800	6,553,500	$3,276,800,000
Period 16	6,553,600	13,107,100	$6,553,600,000

At this rate, how long can Ponzi be expected to continue? Table 8.1 shows his projected schedule of reimbursements at the end of each period, together with the number of new investors he'll need to find during that period and the total number of investors he will have had up to that point.

At the end of a year and three months, he would need to be prepared to pay out $3,200,000; at the end of a year and a half, $6,400,000; after a year and nine months, $12,800,000; and after two years, $25,600,000. Continuing to double every ninety days, we find that after a third year he would need to be prepared to pay out $409,600,000, and after four years, $6,553,600,000.

Given that 1 billion dollars is the total sum of money that John D. Rockefeller (1839–1937), the richest man in American history, amassed over his entire lifetime (the equivalent today is about 200 billion dollars), it is unthinkable that Ponzi could obtain 6 billion dollars in the short space of four years by any method. It is clear that Ponzi will face a major problem in a much shorter time.

2) Investor-Based Model

As it happens, Ponzi had one advantage that gave him a head start on the fast-growing model above: his extraordinary capacity for inspiring hope and trust using his silver tongue. From day one his charm and persuasiveness helped the number of his investors grow at an incredible rate, and as he began paying the promised dividends to those who chose not to leave capital and interest in his hands, the news spread like wildfire. Widespread interest was aroused by the agents Ponzi hired to circulate through the Boston area spreading an unbelievable tale of profit and bringing in ever more money. Indeed, for several months Ponzi actually managed to double the number of his investors each month, which is much faster than the model above, in which the number of investors needed to double every ninety days. Of course, this fast increase in the number of investors brought with it the terrifying necessity of paying out that much more money to all those people.

Let us make another model of Ponzi's scheme, this time based on the doubling of the number of investors every month. As before, we'll start with 100 investors putting in $1,000 each, and assume an investment of $1,000 from every new investor.

TABLE 8.2   INVESTOR-BASEDMODEL

Month	Investors	Amount Received
January	100	$100,000
February	200	$200,000
March	400	$400,000
April	800	$800,000
May	1,600	$1,600,000
June	3,200	$3,200,000
July	4,200	$4,200,000
	10,500	$10,500,000

The progress of the scheme under these conditions is shown in table 8.2, starting in January 1920, Ponzi's first month of operations, and ending in late July, when he ran into trouble. Instead of showing what Ponzi needed to pay out at the end of each month, we show the total amount that he took in, regardless of his expenditures.

This simple model is astoundingly close to what really happened. Adding up the numbers in the table yields a total of 10,500 investors and a total investment of a corresponding $10,500,000. Ponzi's actual records showed that 10,550 investors entrusted him with their money from January through July 1920, for a total of $9,800,000, and that by July he was taking in $200,000 a day, which, amazingly, averages out to exactly $4,200,000 in the twenty-one days of July (not counting Sundays) that passed before the Ponzi scheme collapsed.

If the scheme had not failed—the story is worth telling—it could have been projected into the following months along the same lines (table 8.3):

TABLE 8.3   PROJECTED INVESTOR-BASED MODEL

Month	Investors	Amount Received
July	6,400	$6,400,000
August	12,800	$12,800,000
September	25,600	$25,600,000
October	51,200	$51,200,000
November	102,400	$102,400,000
December	204,800	$204,800,000
January	409,600	$409,600,000
February	819,200	$819,200,000

At the speed at which Ponzi was recruiting new investors, it would have taken him a little over a year to reach the billion-dollar point, and the number of investors would have to correspond to the entire adult population of the Boston area in 1920. At this rate his scheme had no chance of lasting three years as in the more minimal model given earlier; indeed, it could barely have lasted even a few months longer than it actually did.

If investors had made this kind of simple calculation from the start, Ponzi's operation would never have gotten off the ground. As it happens, however, it was cut short before he ever reached the point of being unable to reimburse investors, by the actions of one angry and greedy man.

A FURNITURE salesman named Joseph Daniels had loaned Ponzi $200 in December 1919, just before the whole moneymaking frenzy started. Ponzi had used part of the money to buy some furniture and the other part as investment in his new company. He paid back the debt punctually, but as the money began flowing in over the course of the ensuing months, Daniels decided to claim that Ponzi had promised him a half share in all the profits from the investment. When Ponzi bought his luxury mansion in Lexington on May 28, Daniels paid him a visit and demanded that Ponzi give him a share of the profits. The request was so outrageous that Ponzi refused, upon which Daniels hired a lawyer and sued him on July 2. The district attorney of Massachusetts, acting as a mediator, requested that Ponzi halt operations while an auditor went over his books to confirm that his company was sound.

If Ponzi had cared more about profit than anything else, he would have absconded with everything he held that same day. However, he stayed put. Perhaps he dared not reveal the truth to his wife, or perhaps he believed that his charm would get him through. In any case, he not only remained, but he also kept his doors wide open and a smile on his face.

On Monday, July 26, Ponzi was obliged by court order to suspend operations, which meant turning away dozens of investors who came to him, money in hand. At that time he was taking in nearly $200,000 a day. The next day, however, the news of his frozen operations led to a run on his offices, as investors, frightened by the news of an investigation, came to demand their money. Ponzi made it known that he would redeem unmatured notes in the amount of the original investment and matured notes with the promised returns. Faced with crowds of people who smashed his windowpanes and tried to force their way into his offices, he took over a nearby bar for the day, turned it into a makeshift office, set up a cashier's booth in one corner, and managed to persuade the applicants to form a line. That day some 1,000 clients managed to get their money back, for an estimated total of $1,000,000. When the day's rush was over, Ponzi entertained a group of journalists and shared with them his grandiose plans for the future: he would run for office, make Boston into the largest import-export center in the world, donate millions of dollars to charity, and reform the banking system. The reporters ate it up.

The following day the rush was larger than ever, with thousands of people crowding the street. Ponzi had coffee and sandwiches brought out, and with his best smile he reassured the investors. Anyone who wanted out could have their money at once, but those who believed in him would do better to wait, because all would be well. Many believed him and went home. Ponzi paid the others in full, again for an estimated $1,000,000. On Thursday he paid out half a million, and on Friday only a small sum was paid. His words and actions had calmed the remaining investors and reestablished their belief in him. In the meantime Daniels' lawsuit proceeded.

On Friday and Saturday of that week all was quiet, and on Sunday, August 1, Ponzi attended a fair organized in the nearby town of Jamaica Plain in support of the Italian Children's Home. There, standing on the steps of the home, Ponzi promised the charity $100,000—and paid for every woman and child at the fair to have a free ice cream, a gesture that earned him a rousing ovation as his limousine pulled away (and for which he is remembered affectionately in Jamaica Plain to this day).

On Monday, August 2, the newspapers published an article authored by a former publicity agent for Ponzi, claiming that Ponzi would soon be unable to continue reimbursing clients. This produced another giant run on the offices, which lasted through Wednesday. Once again Ponzi was able to meet demands, although by then he was borrowing from his bank to do it. On Thursday, August 5, things quieted down. But Ponzi knew that if anything provoked a third run, he would be in serious trouble, as the lawsuit had caused $500,000 worth of his assets to be frozen, unusable for paying creditors. He therefore called for a meeting with Daniels to come to an agreement. Daniels agreed to settle for $40,000—it was nothing less than blackmail, as he knew that Ponzi was strapped for cash—and Ponzi handed it over in return for having his assets unfrozen. Unfortunately for him, the whole story filtered into the news, and Ponzi could not avoid another run on Monday, August 9. Once again, he honored his debts. The Hanover Trust Company allowed him to overdraw his account to the tune of $500,000. "I am sick and tired of the whole business," Ponzi is said to have declared on that day.

Tuesday morning Ponzi closed his office and went to give a lunchtime lecture before the Kiwanis Club that had been arranged long before. An enormous crowd attended and asked a host of questions, not all of them easy. People were beginning to wonder about Ponzi's past. On Wednesday the truth about his criminal record was made public, as was his inability to show assets to cover his liabilities.

Ponzi was arrested on Thursday, August 12. He called his wife to tell her that he would be spending the night going over his records with the auditors. She received the guests they had invited for that evening alone, smiling

PONZI CREDITORS CLAMOR FOR MONEY

Crowd Thronging Boston State House Threatens Him and Rival Operators.

PRIVATE BANK IS CLOSED

Commissioner Begins Inquiry on Reports That Policemen Were Ponzi Agents and "Investors."

Article from the New York Times, *August 15, 1920*

courageously. She was the only person there who did not know that her husband was already in prison.

THE RUN continued over the following days until Ponzi's investors finally realized the truth: nothing further was forthcoming. Everything Ponzi owned was seized and sold to satisfy at least some part of his obligation to his creditors. It provides a small measure of satisfaction to note that the $40,000 extorted by Joseph Daniels was also seized and used to fulfill more legitimate claims. Nevertheless, as is necessarily the case in any Ponzi scheme, the greatest number of investors lost their money for good.

Utterly bankrupt, Ponzi was condemned by a federal grand jury to five years in prison for mail fraud. Humiliatingly, he was further sentenced to seven years of prison by the state of Massachusetts, which pronounced him a "common and notorious thief." Because it took such a long time to get around to his trial, and then three tries on different counts before a jury finally found him guilty, at this point it was 1925, and Ponzi had already completed his first prison sentence. He appealed his second sentence at once—one cannot help thinking that it was the adjective "common" that he was appealing more than the accusation of being a thief—and was released while the appeal was pending.

Ponzi had learned his lesson, and in no time he was far from Boston, organizing yet another get-rich-quick scheme to earn money on sketchy land deals in a marshy area of Florida. This time his speculation turned sour quickly, and he and his wife were arrested. On April 21, 1926, he was sentenced to one year of hard labor. Charges against Rose Ponzi were dismissed. Ponzi appealed again, but the sentence was upheld on May 28, by which point, yet again, Ponzi had fled. Tired of the United States—or too frightened to stay—he had managed to find employment on a ship bound for Italy.

Unfortunately for Ponzi, the ship stopped at Port Houston, Texas, where he somehow attracted the notice of the authorities. The Texas sheriff recognized him and quickly wired Boston for his Bertillon measurements (the system of corporal measurements used to identify criminals at the time; more on this in chapter 10, in which this same Bertillon plays a remarkable role). But by the time they arrived, the ship had left Texas for New Orleans. The sheriff immediately contacted a colleague in New Orleans, who inveigled Ponzi off the boat on the grounds of needing standard paperwork, captured him, and brought him straight back to Houston. Ponzi complained that his arrest was no more than an unlawful kidnapping, but he was ignored and Massachusetts demanded his extradition. It took months, but Ponzi was eventually extradited and began serving his Massachusetts prison term in 1927. Released in 1933, he was deported to Italy in 1934, whence he traveled to Brazil in 1939 and held various jobs before a stroke left him paralyzed and penniless. He died early in 1949. The obituary printed by *Time* magazine on January 31 of that year sums up his personality quite well:

> As the dapper little man with the straw hat, the walking stick and the boutonniere emerged from Boston's State House, a cheer went up for "the greatest Italian of them all." Charles ("Get-Rich-Quick") Ponzi shrugged off the compliment. "No," he admitted, "Columbus and Marconi were greater. Columbus discovered America, Marconi discovered the wireless." Hysterical voice from the crowd: "But you discovered money!"

AS THE recent case of Bernie Madoff revealed, Ponzi schemes still have the power to lure unsuspecting—but greedy—investors. And not just one, but thousands and thousands of people; otherwise the plan would never get off the ground. But how is it possible that people did not learn their lesson from the crash of Ponzi's legendary scheme? There are two answers to this question, and one of them comes straight from J. P. Morgan: trust.

The trouble is that, for people like Charles Ponzi—or the Python Kaa in *The Jungle Book*—charisma can produce trust where there should be none. Madoff himself, during the investigation of his own case, noted with amazement that the regulators who investigated his stockbroking company never even asked to look at his stock records. It must be added that Madoff entertained close and affectionate friendships with most of the people in important positions at the US Securities and Exchange Commission who were responsible for such investigations. Between 1992 and 2008, when Madoff was arrested, the SEC was responsible for no less than six different investigations of Madoff, not one of which noticed any anomaly. However, no one has accused the SEC of a cover-up. It is accepted that Madoff's charisma was so overwhelming that investigators could not do otherwise but place their trust in him, exactly as his clients did when they placed their money in his hands. And the investigators continued to do so for years, in spite of the strident cries of a few—there are always some people who simply cannot be hypnotized—such as financial analyst Harry Markopolos, who repeatedly told the SEC that Madoff's returns were mathematically impossible.

And this leads us directly to the second answer to our question above. Ponzi schemes still exist—and work—because not enough people are sufficiently mathematically aware to make the deduction that Harry Markopolos made. As he recounts in his colorful book, *No One Would Listen,* Markopolos tried to alert the industry, the government, and the press about the problem for a period of over ten years. The title of his book succinctly expresses the reactions he received.

Madoff was not arrested until the day his scheme collapsed, as it was mathematically certain to do, and he could not meet the payments he had promised to clients. That day he confessed the truth to his sons. Shocked, they reported him to the police. He was arrested at once, tried, and sentenced to 150 years in prison, where he now hobnobs with such celebrities as Israeli-American spy Jonathan Pollard and crime boss Carmine Persico. An estimated total of 18 billion dollars was definitively lost to his investors, a large number of which were charitable institutions devoted to the support of education, youth groups, and hospitals.

The public failed to learn the lesson of the Ponzi scheme from its namesake. Perhaps the experience with Bernie Madoff can have at least one positive consequence: public understanding of the mathematical impossibility of such terms.

Beware exponential growth investment—it cannot work!

MATH ERROR NUMBER 9 »
CHOOSING A WRONG MODEL

A farmer is troubled because his hens are not laying eggs. After visiting a veterinarian, a doctor, and a psychotherapist with no result, he ends up going to a physicist in despair. The physicist scratches his head and says, "Give me a week to think about it."

A few days later the anxious farmer phones the physicist. An enthusiastic voice shouts into his ear, "Yes! I have found a solution to your problem! The thing is this, though. My solution only works for perfectly spherical hens in a vacuum."

—ADAPTED FROM *THE BIG BANG THEORY*

IF YOU LAUGHED at this little story you probably understand that the funniness of the joke lies in the contrast between a purely scientific situation and the complexity of real life. It's amazing how often scientists fall into the trap of believing that their models accurately describe real situations when in fact everyone knows that reality can be stranger than even the wildest fiction.

Applying a mathematical model to a real-life situation is never likely to be completely accurate; the simpler the model and the more individual the behavior, the worse it will turn out. It should go without saying that this kind of reasoning cannot be used in a court of law without incurring a serious danger of being wrong.

The Case of Hetty Green:
A Battle of Wills

Hetty Green's life was not what you could call ordinary. The only surviving child of a hard-nosed, extremely successful businessman, she grew up under her father's wing, learning the ABCs of money management before age ten by reading the financial news to him every day. Instead of dolls, stocks and bonds were her playthings. Her bulls and bears were not stuffed ones, and legend has it that upon receiving her first cash gift at the age of eight, she rushed to the town bank on her own to open an account. Like her father, Hetty grew up stubborn and thrifty—the amassing and later the multiplying of money became her passion, her obsession, and finally the only activity of her life. Known in old age as "the witch of Wall Street," she was the richest woman in America and the only female investor on Wall Street, with railroads, factories, and entire city blocks in her hands. More than once she bailed out the entire city of New York by offering a gigantic loan to shore up the city's failing budget. During the famous Panic of 1907, she came to the city's aid with a loan of $1.1 million from her personal fortune. When she died in 1916, she was worth $100 million (estimated at $1.6 billion today), the thirty-sixth largest American fortune of all time. Hetty Green, née Robinson, was a legend in her own time, a legend fueled by innumerable anecdotes about her peculiar behavior anywhere the spending of even a penny was concerned.

In 1865, Hetty was thirty years old, engaged to be married to a wealthy man, and possessed of a generous fortune left to her by her father on his death shortly earlier: a million dollars in cash and real estate and another five million in trust, with the interest to be paid to her by the trustees over her lifetime. One might think she had everything a young nineteenth-century woman could possibly want, but at that point in her life, Hetty was a very angry and frustrated person. The source of her anger was her father's will, which, far from the boon it might appear to be, struck her more like an insult or a slap in the face. Being deprived of control over her own money was a humiliation that Hetty found unendurable. Her father had raised her and had shared his financial knowledge with her practically since her birth. He knew her passion for the management of money; he knew her care, her thrift, her frugal habits, her intelligence, and her instinct. He knew who she was. And yet he dared to leave her money— *her five million dollars*—in trust, handled and managed by men, as though she were some foolish girl who might spend it all on ribbons and parties. Hetty was deeply wounded by her father's will, and she never forgave him for it.

Hetty Green in her old age

AT THE time of her father's death, Hetty was expecting a second inheritance: that of her mother's sister, Sylvia Howland. Hetty's mother had died when Hetty was still a child, and the rambunctious, temperamental, strong-willed young girl had spent a great deal of time at the house of her invalid maiden aunt in New Bedford, Massachusetts. Sylvia's fortune was reputed to be immense—Hetty's father had married into money—and Hetty had been brought up with the expectation that she would inherit her aunt's money, since there were no other close relations. Determined that the misfortune of her father's legacy would not be repeated, and seeing her aunt constantly surrounded by companions, nurses, and caretakers, Hetty undertook to ensure that her inheritance would not be altered or diminished by any untoward fancy of the old lady, and decided to convince her aunt to make a will.

Poor Sylvia was reluctant to do so, according to the testimony of her companions. It was not that she didn't want to leave her money to Hetty, but she was disturbed by her niece's overbearing ways, her inelegant insistence on the matter of this will, her emerging eccentricities of stinginess and personal negligence, and her temper, which made her fly into a rage and reproach her aunt bitterly for such things as wishing to build an addition to her house. And perhaps Sylvia also disliked the idea of writing her will with Hetty standing over her, breathing down her neck, frowning and throwing

tantrums at every mention of a legacy left to someone else, no matter how
dear to Sylvia Howland's heart. In any case, Sylvia had already made a will
that satisfied her, by which Hetty was to receive two-thirds of her estate in
trust, with the remaining third divided among legacies and charities that
Sylvia cared about.

But this proposal was not acceptable to Hetty. A third of the estate passing out of family hands horrified her, the idea of a trust even more. She
began to pressure her aunt to rewrite the will. Sylvia initially resisted, but
Hetty insisted and they eventually made a bargain: they would both write
wills. Sylvia would write a will in Hetty's favor, and Hetty would write a
will leaving half of her own money to her children, should she have any,
and the rest (or all of it if she remained childless) to Aunt Sylvia's favorite
causes, the local children's home in particular. Young and strong, Hetty was
well aware that she had nothing to lose; her will was a mere gesture of good-
will to persuade her aunt to make another in return. Sylvia yielded, probably
out of weariness, and "dictated" a new will to Hetty—one can imagine how
that dictation proceeded—according to the terms of which Hetty inherited
every single bit of her aunt's estate, houses, land, money, investments, and
holdings.

Once the will was written, however, Sylvia refused to sign it. A tense
standoff ensued, with Hetty fuming, Sylvia withdrawn, and the nurses and
servants embarrassed. Hetty declared that she would not return home to
New York until the will was signed and witnessed. Thus the bedridden patient's only pleasure—a life of peaceful, quiet enjoyment with her caretakers,
friends, and visitors—was denied her by her intransigent niece. In the end,
Sylvia's desire to return to her normal habits overcame her resistance. She
called for Hetty and for two trusted friends, and the will was signed and witnessed with all due formality. It was then put into an envelope and placed
in a trunk containing Sylvia's private affairs, to which only she and her housekeeper had the key. Hetty, relieved, left for New York.

Despite everything, Hetty remained unsettled, because it was obvious
to her how lonely her aunt was and how dependent she was on those who
called themselves her dependents. Sylvia could write another will anytime
she chose, and there wouldn't be much that Hetty could do about it. And
she must have been aware that having forced Sylvia's hand so rudely, the
event she feared was all the more likely to occur.

There wasn't much for Hetty to do, however, except to hope for the best and keep her ears open. She went back to New York, became engaged to Edward Green, and wrote conciliatory, affectionate letters to her aunt. And she waited.

IT WAS not long before a worrisome rumor reached her ears. There was a new influence in Sylvia's life. A doctor named William Gordon, who had arrived recently in New Bedford, had been called in by Sylvia to treat her chronic back pain and had by degrees become her regular and then her constant doctor. In fact, she took up so much of his time and attention that she eventually became his only patient, with him spending most of the day at her bedside. Dr. Gordon brought Sylvia much relief, but there was a steep price to pay, as the painkiller he prescribed was the popular nineteenth-century opium-based drug laudanum, which kept her in a feeble, drowsy, and addicted state. Sylvia's health was in rapid decline, and the doctor's influence over her reached a point that appeared excessive to everyone around her.

News of the doctor's rise in fortune reached Hetty and caused her tremendous anxiety. What pushed her over the edge was the receipt of a letter from the same Dr. Gordon, informing her that according to her aunt's wishes, Hetty was no longer welcome to visit the house. Because this maneuver effectively tied her hands, she was compelled to stew in New York, virtually certain that all the while she was being robbed of the fortune that she considered rightfully hers, and that it was being diverted to unworthy and scheming hands.

It must have been a difficult time for Hetty, and she can be forgiven for indulging in numerous plans destined to foil her aunt's project. But from a distance of over one hundred miles, even the most willful and headstrong young woman cannot have the same influence on a failing invalid as a helpful and caring doctor who is present at her bedside every single day. Nearly two years after her deal with Hetty, Sylvia made a new will, one that was calculated to send her niece mad with rage.

Of the two million dollars Sylvia's estate was worth altogether, she bequeathed fully one-half to a collection of charities and individuals she wished to benefit—above all, the town of New Bedford and members of her family, together with her faithful companions, housekeeper, and nurse, all of whom received legacies of between three thousand and twenty thousand dollars.

The remaining million dollars was left to Hetty, but in the way her niece would most abhor: bound up in a trust, administered by *yet more men*—one of whom was the detested Dr. Gordon himself. The trustees were given complete freedom to control Hetty's money; indeed, though they were directed in Sylvia's will to pay the net income from their investments to Hetty, they were not bound by any express stipulation to actually do so, as she wrote: "I wish said Trustees to make said payments when and as often as it may in their judgment be convenient for them to do so . . . It is my will that the said Trustees shall exercise their own judgment, and shall act and do in all respects what shall be deemed by them to be for the interest of all parties . . ." And the cherry on the cake was that, not content with providing for generous and permanent trustees' fees to ensure the helpful doctor's future, Sylvia left him outright the sum of fifty thousand dollars! The will was dictated to and written down entirely by the good doctor himself, who all the while was administering to his patient increasing doses of laudanum.

Hetty, of course, was kept in the dark about all these changes, but when Sylvia died six months later and the complete state of affairs was revealed to Hetty's shocked ears, she must have seen the signs of what is now known as "undue influence." As with her father's will, she felt that her aunt's will was a betrayal. And Hetty resolved, come what may, to contest the document in court.

Hetty was no fool, and she immediately realized that she could not contest the will on the grounds of Sylvia's feeble mental state. She had not seen her aunt for over two years, and all of Sylvia's friends would testify that she was in full possession of all her faculties. Even those who might have felt that Dr. Gordon had reaped more than his due were unlikely to protest a will that also carried benefits for them. Hetty wanted everything for herself, and she knew well that she would have no supporters for the position she was going to take—no supporters, at least, aside from her fiancé, Edward Green, and a few carefully chosen and well-paid lawyers. If she wanted to contest the will, she would have to do it on grounds that were more solid than mental incompetence. The best angle would be the mutual promises that she and Sylvia had made to each other, proven by their cosigned wills. But in order to support this position, it was first necessary for Hetty to obtain the will that she herself had written down at Sylvia's dictation. At that time it was not as common as it is now to leave one's last will and testament in a bank vault or in the care of a lawyer. Sylvia had not written that first will with the help of a lawyer, so Hetty suspected it would probably still be in the house. If it was, she knew where to find it: in Sylvia's special trunk, the one to which only Sylvia and her housekeeper held the key.

On the very evening of Sylvia Howland's funeral, Hetty, holding a candle and accompanied by the housekeeper and the nurse, went up the dark stairs to Sylvia's room, had the trunk opened, and retrieved some papers that were inside: two large envelopes, one containing her own will, and the other, the one dictated to her by her aunt. Both the housekeeper and the nurse later testified to seeing Hetty remove these papers, although they did not know what they were. But Hetty must have been excited to find that her aunt had not destroyed the previous will. On the strength of the document, she contacted the probate court judge who would decide whether to admit Sylvia's will, and presented her strongest arguments against it (not excluding a hint of bribery). The judge ignored them, however, and admitted the newer will. Hetty decided to sue the trustees.

In her suit Hetty claimed that as the sole legitimate heir to the estate, she was entitled to inherit it in its entirety. She accused Dr. Gordon of drugging and manipulating a feeble, elderly lady for his own benefit. When not under his pernicious influence, she claimed, Sylvia had meant Hetty to inherit the entire estate, as the earlier will clearly proved. The trustees countered with a response, filled with witness statements, pointing out that during the last years of her life Sylvia had changed her mind about her niece because Hetty was so pushy and unpleasant. They cited Sylvia's nurse as saying that when Hetty used to visit, Sylvia would beg not to be left alone with her; other members of the household staff came forth with a plethora of unpleasant anecdotes in which Hetty was described in such embarrassing postures as screaming and rolling on the floor or pushing servants violently down the stairs. The document described Sylvia's relief at being able to dispose of her money as she wished, providing all the kindnesses she had always desired to the children's home, to local widows, and to her friends. The trustees agreed that Sylvia's will was absolutely sound and unbreakable. Thus the ball was back in Hetty's court, and she swept it into the air with a backhand as astonishing as it was unexpected.

There was a second page attached to her aunt's old will, Hetty explained, which had been dictated to her by Sylvia on the same day as the will itself, in secret. The witnesses had not seen it, and it had been signed by Sylvia alone. There was nothing surprising in that, given the contents, which were intended to be kept from everyone except Hetty. Even at that time, Hetty contended, Sylvia was aware of the dangers of exploitation that surrounded her, a wealthy woman dependent on others for her care, and she had written this letter in order to protect herself from possible quarrels with her carers over questions of inheritance. The letter automatically invalidated any future will that she might write.

> Be it remembered that I, Sylvia Ann Howland, of New Bedford, in the County of Bristol, do hereby make, publish and declare this the second page of this will and testament made on the eleventh of January in manner following, to wit: Hereby revoking all wills made by me before or after this one—I give this will to my niece to shew if there appears a will made without notifying her, and without returning her will to her through Thomas Mandell as I have promised to do. I implore the Judge to decide in favor of this will, as nothing would induce me to make a will unfavorable to my niece, but being ill and afraid if any of my care-takers insisted on my making a will to refuse, as they might leave or be angry, and knowing my niece had this will to show—my niece fearing also after she went away—I hearing but one side, might feel hurt at what they might say of her, as they tried to make trouble by not telling the truth to me, when she was here even herself. I give this will to my niece to shew if absolutely necessary, to have it, to appear against another will found after my death. I wish her to shew this will, made when I am in good health for me, and my old torn will made on the fourth of March, in the year of our Lord one thousand eight hundred and fifty, to show also as proof that it has been my lifetime wish for her to have my property. I therefore give my property to my niece as freely as my father gave it to me. I have promised him once, and my sister a number of times, to give it all to her, all excepting about one hundred thousand dollars in presents to my friends and relations.

The letter, dated January 11, 1862, was written out in Hetty's handwriting but signed by Sylvia. Hetty presented it to the court. The trustees learned about it, looked at it, and pronounced it a fake. And the battle began.

FOR A woman whose entire life had dealt with numbers and calculations, forecasts and predictions, and statistically informed investments, it is striking that Hetty Green's major brush with the arcana of probability theory occurred not within the framework of monetary activities, but in a different arena altogether: the trustees' accusation of forgery.

Thomas Mandell, the executor of Sylvia Howland's will, challenged Hetty's story, as well as the validity of the earlier will altogether. Mandell examined the signatures on the earlier will—the one on the main page of legacies and the ones on the additional document (in two copies) deeming all future versions worthless—and found all three to be virtually identical. Accusing Hetty of having forged the second page and its signature in order to inherit the whole of her aunt's fortune, he stood up to her in court.

It is hard to imagine a trial taking place today in which neither side, uplifted as they were by a sense of wealth and entitlement, hesitated to call upon the most important academics of the day as expert witnesses on a question as trivial as that of a forged signature. But unlike most other accused forgers, Hetty's privileged family background gave her access to the East Coast's intellectual elite.

Her defense opened by claiming that near-identical signatures are not as rare as one might think. To prove it they called in an engraver, J. C. Crossman, who testified to the effect that many people have signatures of astounding regularity. As an example, he provided several specimens of the signatures of former president John Quincy Adams, some virtually identical to each other. Next, the defense sought to prove that Sylvia Howland's signature on the addendum letter was not forged. For this they called upon no less an authority than Louis Agassiz, professor of natural history at Harvard and the first scientist ever to theorize the existence of an Ice Age. Agassiz examined the purportedly forged signatures under a microscope and declared that he could see

Louis Agassiz, naturalist

Oliver Wendell Holmes

no signs of forging, none of the minuscule tremors that appear in inked lines that have been drawn carefully and slowly, or traced, rather than put down with a bold, firm stroke.

Finally, to disprove the theory that Hetty might have first traced Sylvia's signature with a pencil and subsequently gone over it with a pen, Hetty's legal team called upon Oliver Wendell Holmes, Parkman Professor of Anatomy and Physiology at the Harvard Medical School. Holmes was one of the most famous men in America, not only for the popular poetry and essays he published in the *Atlantic Monthly* but also for his medical innovations. Not only was he one of the first proponents of the use of anesthesia during operations, but he even coined the word itself, predicting that it would soon "be repeated by the tongues of every civilized race of mankind."* Ahead of his time both scientifically and socially, he was also famous (or infamous) for trying to admit the first woman applicant to Harvard Medical School in 1847. The united opposition of the student body, the university overseers, and the other faculty members thwarted his attempt to do so, but three years later he managed to briefly admit three black men to the school. Alas, a petition was circulated and signed by more than half the student body, saying, "We have no objection to the education and evaluation of blacks but do decidedly remonstrate against their presence in College with us," and once again Holmes was obliged to bow to public pressure and shorten the black students' education to a single semester.

---

*On top of this, Holmes is still known today as the discoverer of the fact that the puerperal fever, which caused the death of thousands of new mothers in hospitals all over the country, was spread from one mother to another by the hands of the physician attending to her. It is thanks to Holmes that doctors now routinely sterilize their hands, clothing, and instruments. This theory caused considerable furor in the obstetric milieu when Holmes first published it, as many doctors were angry about being accused of having killed their patients through lack of hygiene. But Holmes, having studied hundreds of cases of puerperal fever and its spread, knew that he was right and implored the world to heed his words: "I beg to be heard on behalf of the women whose lives are at stake, until some stronger voice shall plead for them."

The illustrious but controversial Oliver Wendell Holmes agreed to make a detailed scientific examination of the signatures that Hetty Green was accused of having forged, using the best possible equipment his laboratory at Harvard could provide. He rendered his personal opinion as a witness that they showed neither pencil marks nor any other of the typical signs of tracing.

For the prosecution, this provided something of a puzzle. There was obviously no easy way to prove that the signatures were forged in the face of such illustrious scientists claiming the contrary. Thus they came up with a new strategy: instead of calling upon specialists in the microscopic biological sciences, they decided to counter the scientific testimony with mathematics. For this purpose they also went straight to the top, consulting the professional opinion of Benjamin Peirce, a professor of mathematics at Harvard (whose name is attached today to the prestigious Benjamin Peirce lectureships there) and his son Charles Sanders Peirce, a renowned logician and philosopher. Benjamin Peirce undertook to test whether or not the disputed signatures were forgeries, using a most original method.

THE IDEA in itself appears simple and quite convincing. Under instructions from his father, Charles took 42 examples of Sylvia Ann Howland's signature from documents found on her estate, laid every single signature over every single other one for a total of 861 comparisons of pairs, and in each case observed how similar the two signatures were.

Charles Sanders Peirce

As a measure of similarity, he chose to count the number of downstrokes that were superimposed on each other. In order to analyze the lettering in detail, he counted as many downstrokes as possible, including the tiny loops at the start of each capital letter, and came to a total of 30: 9 in the word "Sylvia," 7 in the word "Ann," and 14 in "Howland" with its complicated capital *H*.

Next, he made a table showing how many of the 861 pairs had 1 downstroke in common, how many 2 downstrokes, how many 3, and so forth, up to 30 strokes

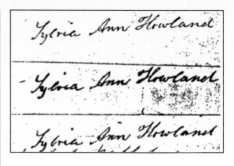

(Left) Some of the 42 signature samples
of Sylvia Ann Howland measured by
C. S. Peirce

(Above) The signature on the first
page of the will and the two
disputed signatures

in common. The table that remains in the trial transcripts, given in the first
and second columns of table 9.1, shows a bit of abbreviation in what concerns the very similar signatures, lumping all those with 13 to 30 downstrokes in common into a single group.

This table indicates that most of the pairs, or 617 out of the total 861,
had 3 to 7 downstrokes in common, and only 20 had more than 12 downstrokes in common. So how did the two Peirces use this table to calculate
the probability that some pair might actually have all 30 downstrokes in
common?

In court, Benjamin Peirce gave an estimation of that probability as a
chance of 1 in 5^{30}, which he qualified as unspeakably small. He told the
court that this number was equal to 1 in 2,666 millions of millions of millions (which is actually about three times too large—but there were no cal-
culators in the nineteenth century). In any case, the words he used to
describe the quantity $1/5^{30}$ apply just as well to the correct value: it was a
"number far transcending human experience . . . so vast an improbability is
practically an impossibility. Such evanescent shadows of probability cannot

TABLE 9.1

Downstrokes in Common	Number of Measured Pairs	Theoretical Model
0	0	1
1	0	8
2	15	29
3	97	68
4	131	114
5	147	148
6	143	155
7	99	132
8	88	95
9	55	58
10	34	31
11	17	14
12	15	5
13–30	20	3
Total	861	861

belong to actual life. They are unimaginably less than those least things which the law cares not for."

What he meant, of course, and what the jury took him to mean, was that the probability that the two signatures were identical by pure chance was so tiny as to be utterly negligible, and therefore they must be identical by design. And whose design could that be, if not that of the interested party, Hetty herself? In other words, Peirce's statement was tantamount to a straightforward but unspoken accusation of forgery. But this statement, and the calculation in general, deserve closer attention. Why, if there were 20 signatures having 13 to 30 downstrokes in common, was the probability he calculated for two signatures having 30 common downstrokes so incredibly tiny? And for that matter, how did he calculate that surprising figure at all?

What Peirce did was to approximate the numbers in the table of measured values using a simple standard mathematical model, the binomial formula.* The numbers predicted by the binomial formula are in the third

---

*This formula is given by $861 \times \binom{30}{n}(1/5)^n(4/5)^{(30-n)}$, where $\binom{30}{n}$ is the binomial coefficient for each number n between 0 and 30.

column, titled "Theoretical Model," of table 9.1. A brief glance shows why Peirce picked that model: the numbers in the third column and those in the second column seem fairly close to each other, which means that the model seemed like a good choice.

But a closer look shows that the numbers are not as similar as they seem at first glance. To make a more accurate comparison, let's put the two tables of values in the form of a graph: white for the actual measurements, and black for the binomial model.

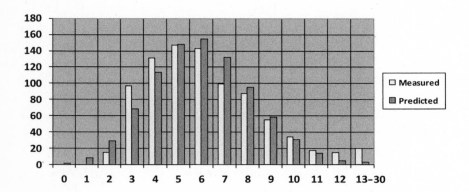

Although the black and white graphs have similar shapes, they are actually quite different from each other at both ends. What do these differences mean?

The absence or smallness of the white (real) bars compared to the black (theoretical model) bars on the left side of the graph indicates that Sylvia Howland's signatures were generally more similar to one another than a purely random distribution would predict. Next, the greater height of the white bars compared to the black ones on the right edge of the graph shows that Peirce found 20 pairs with 13 to 30 downstrokes in common when the model only predicted 3. And for pairs having more than 11 strokes in common, the model (black) predicts only 8 altogether, whereas Peirce's measurements (white) revealed 35 such pairs, more than four times as many.

It is curious that the number of signatures having 13–30 strokes in common were not specified in the Peirces' table. Of course, if even one pair were to be found with a large number in common, such as 27, 28, or 29 strokes, the probability of such an event happening would be shown to be at least

large enough to justify the fact that it just might have happened in the particular case at court—so certainly not anything as small as the "vanishing"
$(1/5)^{30}$. It would have been interesting to know if they had actually found
any such pair or not.

In any event, comparison of the theoretical model with the reality shows
that not only is the correspondence far from perfect, but it also underesti-
mates the similarities that Sylvia Howland tended to have among signatures.
It was therefore quite daring on the part of Benjamin Peirce to conclude by
a $(1/5)^{30}$ probability of two signatures being identical on that basis. In fact,
with all due respect for the great man, the figure he gave and the comments
he made about it are downright absurd.

For one thing, the model that gave rise to the transcendentally small
probability takes no account of a number of factors that could explain the
large number of similar pairs. Sylvia's signature could have slowly changed
over time, so signatures made soon after one another would be very similar,
whereas those far apart in time might be different. Also, signatures made
using the same pen, at almost the same moment, sitting in the same position, might be expected to be more similar than those made under different
physical conditions. A model taking these considerations into account would
have yielded a much larger probability for identical signatures. According to
Hetty, the signatures on the will and on the addendum letter were made one
right after the other, with Sylvia presumably not moving between the two.
This would naturally yield signatures more similar to each other than to
other signatures made months or years earlier.

There is also the matter of the interpretation of Peirce's vanishingly tiny
figure. Because he calculated this number as the probability that the two
signatures were identical purely by chance, he concluded that such a thing
was virtually impossible, and therefore the second signature must have
been forged by Hetty. But as statisticians Paul Meier and Sandy Zabell have
humorously pointed out, there are many other possibilities that, however
outrageous, are all far more probable than $(1/5)^{30}$. Sylvia Howland might
have decided to trace her own signature, they suggest. Or Thomas Mandell,
Sylvia's executor, might have stolen the real second page of the will and
forged a copy of the whole thing with a traced signature in order to destroy
Hetty's claim. Or Charles Sanders Peirce might have been convinced of
Hetty's guilt and thus have been particularly severe in his measurements

of similarities of those two particular signatures, compared to the manner in which he measured the others. All of these scenarios (and many others one could reasonably imagine) are unlikely, to be sure. But they are nowhere near as unlikely as Benjamin Peirce's 1 in 2,666 millions of millions of millions!

Yet not a single one of these questions was raised at the trial. The judge simply opted to reject Hetty's testimony altogether, as she was an interested party, and the case ended with a settlement that was essentially identical to Sylvia's latest will. Some members of the family, furious with Hetty for her behavior, threatened to bring her to court over the forgery. Hetty married Edward Green and, probably to avoid another trial, went to live with him in England, where the couple had two children. Therefore it is left to historians to ask themselves whether, had that second trial taken place, she would have been convicted. Our conclusion is that although it is not unlikely that she did forge the signature, the Peirces' calculation certainly did not establish that fact beyond a reasonable doubt.

WHAT IS CERTAIN, though, is that Hetty's frustrating experiences with money played an essential role in forming her into the intransigent, indomitable moneymaking machine that she later became; the story is worth telling. The family did not return to the United States until 1879, when they settled in Green's hometown in Vermont, Bellows Falls. Hetty's husband was a wealthy and successful businessman, but unlike his wife, he enjoyed spending his money and living well. At first, she allowed him to do as he liked on the condition that he never touch her personal fortune, most of which was held in trust for her, with the interest paid out in small installments. Hetty never spent these amounts, but deposited them immediately in the bank, where eventually they grew to form an appreciable sum. During these years, however, she developed increasingly miserly habits, to the point where she ended up furnishing the townspeople with frequent laughs—"Did you hear how Hetty Green awoke her hostler late at night and forced him to spend hours with her in the chill darkness, searching her carriage and stable, and then the lawns and grounds of an inn where she had dined that day, for a two-cent stamp that she lost?"—and seriously annoying her husband. It is told that she rendered him speechless with rage on the day of his mother's funeral, when in answer to a sharp question of why wine was being served to the guests in cheap glassware, Hetty explained that she had put all the expensive crystal glasses away in boxes, as "there was no point risking valuable heirlooms when everyday glasses held the liquid just as well." Contemporary accounts have it that Edward stared at her, dashed his glass straight into the wall, and walked out of the room. By this time Hetty's attitude about money was causing tremendous tensions in her marriage.

Although not deeply religious, Hetty was of Quaker background, and it was to this that she often attributed her desire for a simple life, devoid of the trappings of luxury and external indicators of wealth. But there may have been another reason for her asceticism. Without engaging in amateur psychological analysis, one may note that the following passage from the *Encyclopedia of Mental Disorders* closely resembles the portrait of Hetty that emerges from numerous biographies.

> Obsessive-compulsive personality disorder is a type of personality disorder marked by rigidity, control, perfectionism, and an over-concern with work at the expense of close interpersonal relationships. Persons with this disorder often have trouble relaxing because they are preoccupied with details, rules, and productivity. They are often perceived by others as stubborn, stingy, self-righteous, and uncooperative.

Difficult to live with she surely was, but both parents dearly loved their son Ned and their daughter Sylvia, and even after raising their children to teenage, they might have hobbled along as a couple for many more years had not a catastrophe struck, close to Hetty's fiftieth birthday, in the form of a bank collapse.

Cisco and Son was the New York bank where both Hetty and Edward placed the bulk of their money, which in Hetty's case amounted then to some half a million dollars, not counting the much larger amount of money held in the bank in the form of securities. When rumors that the bank was troubled reached Hetty, she hurried to New York at once and demanded that her entire deposit be transferred to another bank. Although Hetty had every right to request what she did, she was shocked when the bank refused. It was then that she learned for the first time that her husband was not the wealthy man she had believed him to be. He was in debt to Cisco and Son for a whopping

$702,000—and the bank refused to let Hetty's money go until she made good on her husband's debt!

Oh, the humiliations of marriage. In vain Hetty stormed, raged, refused to pay back loans that had not been made to her, and told the bank that she was not responsible for her husband's affairs. In vain she ignored the increasingly pointed series of articles about "Mrs. E. H. Green's behavior" that appeared in New York newspapers, pointing the finger squarely—and not at all fairly—at her as the major cause of the bank's demise. But Cisco and Son, now taken over by another firm, continued to refuse to give her the money until she paid off at least a significant part of her husband's debt.

Over several days in January 1885, visitors to the bank, mostly small depositors desperate to take their savings out before Hetty's demands made doing so impossible, were witnesses to an astonishing scene. Hetty sat at the new bank director's desk and raged. It is said that she also stomped, cried, and even screamed. The director sat calm and unmoved. Passersby in the street gathered at the windows to watch.

Hetty wanted her half-million deposit paid out to her and her securities returned. The bank wanted her to pay back her husband's debt. Unfortunately for Hetty, the main bargaining chip was the fact that the securities were on paper, and that bundle of paper was located in the bank's strongbox. She could not have them until the bank decided she could, and they were not going to decide that until she paid her husband's debt.

Hetty repeated that her husband's finances were no concern of hers. The bank director fiddled with his pen, worked quietly at his desk, and made it clear that he was not in any hurry. Hetty was furious to be unable to get hold of her securities, which deprived her of control over her money, a galling situation that she had already had to endure twice with the trusts inherited from her father and aunt. It was a question of one humiliation weighed against another. With fingers stiff as claws, as one can well imagine, Hetty wrote out a check to the bank for the disputed $702,000, took her bag of securities, and marched out, pale with rage, to place her remaining fortune at a more trusted place, the Chemical National Bank.

As might be expected, her anger was soon turned toward its real cause: her husband. That he should misspend and squander his own fortune was not her affair, and perhaps she would not even have cared much about his losing his own money. But he had committed the unforgivable sin of making a hole in her wealth, and this she could not and would not accept. She took her children, then ages seventeen and fourteen, and moved out of the family home in Bellows Falls.

For the remaining thirty years of her life, Hetty Green lived in a series of modest two-room flats in brick tenements in Hoboken, New Jersey, and other cheap towns that allowed her to reach New York by an inexpensive early morning commute—by public transportation, of course, since she did not feel that she could afford a carriage. Each weekday, she went to the desk that had been placed at her disposal in the main banking room of the Chemical National Bank on Wall Street and spent her working hours selecting and buying her investments—rarely either developing or selling, but buying, buying, buying, and waiting for values to rise. She described most of her work there as "cutting bond coupons with a large pair of scissors." At lunchtime, having rejected most of the surrounding eateries as too expensive ("Ten cents for a cup of tea? It isn't worth it") or annoying the waiters by offering them advice instead of tips, she would often simply add water to a small pail of dry oatmeal she brought in with her from home, and set it on the radiator to heat. The only thing in life on which Hetty seemed willing to actually spend money was lawsuits, directed against those she felt had wronged her, including the director of Cisco and Son who had forced her to write the dreadful check. When questioned—and quite often scolded for interrupting proceedings—by the judge, Hetty would say: "I come of good old Quaker blood. All I care for is to do right. Then I am sure to go to heaven." As the years went by, she became recognizable everywhere, by her reputation, which made people point her out to each other, and by her strange garb, always a long, black dress and a cloak and bonnet, giving rise to the half-admiring, half-frightened epithet "witch of Wall Street."

Her son left home to become a respected and talented investor and moneymaker in his own right, and her daughter got married. Hetty lived alone and worked, never spending, but steadily increasing her fortune, investing in chunks of America itself: whole streets in Chicago, newly developed railways, empty land tracts. As prices rose and fell, she sometimes cleared as much as two hundred thousand dollars in a single day, and unlike Ponzi, she did so both honestly and legally. It is a feat that would be considered spectacular

Green Hall, Wellesley College

even now, 130 years later. Everyone agreed that Hetty had flair, even a kind of genius, for investment opportunities. The money that she had started with, inherited and left to her in trust, hardly increased over all those years due to poor management and ended up being a tiny, almost unnoticeable part of her gigantic fortune, one of the largest that America had ever seen.

When Hetty Green died in 1916, her fortune passed to her son and her daughter, both childless. Upon their deaths, it filtered away in gifts to individuals and associations, universities, libraries, and charities. Wellesley College's Green Hall in Massachusetts is the only public building actually named for Hetty. Partly because her children did not leave many legacies specifically in her name, it has been largely forgotten in comparison with the other unforgettable makers of fortune of her time: Carnegie, Morgan, Rockefeller. But Hetty had always wanted a modest life and probably would prefer it this way.

IN THE 150 years that have elapsed since the trial, it has become pretty widely accepted that Hetty did forge her aunt's signature on the famous second page. But is it true? If she did, one wonders, for example, why

Sylvia would have kept the first will in her house instead of simply destroying it.

When Hetty presented the first will and the famous letter to the court, the two pages were pierced with tiny holes around the edges. Hetty explained that they had been sewn together, face-to-face, so as to discourage the curious, by which she meant Sylvia's companions, nurses, and servants—Hetty's sworn enemies. If that were true, however, she would have been wiser to show these sewn papers to those same sworn enemies at the very moment when she extracted them from Sylvia's old trunk, so that she would have witnesses to their existence. Instead, she carried the papers home to examine them by herself. It is a suspicious circumstance, yet one that corresponds to the controlling, power-hungry element of her personality.

At no time in her life, subsequently, was Hetty ever accused of any kind of cheating or dishonesty. She was known for many colorful character defects, but certainly not that one. Nor did she ever admit to the forgery of which the court's judgment implicitly accused her. For that matter, it is important to note that she was never formally accused or brought to trial for that forgery, not even by Thomas Mandell or Dr. Gordon, the executors of Sylvia's will, who had every reason to execrate her. No one will ever know the truth.

Hetty's subsequent history does show, however, that whether she cheated on a gamble and lost, or whether she was actually defrauded of something that was legally hers, everything that happened worked out for the best, financially. Without overcoming the frustrations and obstacles that forged her character, Hetty Green would probably never have become the legendary witch of Wall Street, female tycoon.

MATH ERROR NUMBER 10 »
MATHEMATICAL MADNESS

LIKE THE "INCREDIBLE COINCIDENCE" in Math Error Number 7, the error addressed in this chapter involves the likelihood of an unlikely event occurring, except that here we deal with the case where an event occurs several times, rather than just once as in the lottery example. This error consists in calculating the probability, for instance, that some relatively rare event will happen ten times, but forgetting that those ten "successes" were the result of one hundred attempts. For example, before being amazed that your friend shot five arrows into the bull's-eye, you might want to count how many arrows he shot into the colored rings and the surrounding wall. The fewer total arrows he shot altogether, the more impressive his performance; if he shot a hundred arrows or more, it becomes less of an achievement.

The website www.anxieties.com has a page devoted to overcoming a fear of flying. While the site's goals are laudable, it is unfortunate that such misleading passages as the following are used to reassure people:

> Dr. Arnold Barnett, of the Massachusetts Institute of Technology, has done extensive research in the field of commercial flight safety. He found that over the fifteen years between 1975 and 1994, the death risk per flight was one in seven million. This statistic is the probability that someone who randomly selected one of the airline's flights over the 19-year study period would be killed in route [sic]. That means that

any time you board a flight on a major carrier in this country, your chance of being in a fatal accident is *one in seven million*. It doesn't matter whether you fly once every few years or every day of the year.

Apart from the facts that the period from 1975 to 1994 does not comprise fifteen years, that the term "the airline" seems to contradict "a major carrier," and that Dr. Barnett's research appears to be valid only for flights originating in the United States, this passage contains a major mathematical flaw. Try replacing "boarding a flight" with "playing Russian roulette" and the probability of "one in seven million" with "one in six." The last part of the passage now reads: "Any time you play Russian roulette, your chance of being in a fatal accident is one in six. It doesn't matter whether you play once every few years or every day of the year." Is "one in six" really the right figure to consider here? Does it really not matter if you play Russian roulette once every few years or every day of the year?

How do you calculate the probability of getting a particular result several times over a given number of tries? Imagine you roll dice against a partner 6 times, and 3 of those times he rolls a six, with the end result that he wins the game and carries off the stakes. Rolling a six 3 times may seem so extra lucky, so unlikely to occur naturally, that you might be tempted to accuse your gaming partner of having a loaded die. Before accusing him of cheating, however, a little calculation is in order. Exactly how unlikely is it? The only way to answer this is to calculate the probability that in 6 rolls of dice, we land exactly 3 times on the number six.

There is a 1/6 chance of landing on six, and one common error would be to conclude that the probability we are looking for is $(1/6)^3 = 1/216$. Such an error is what makes coincidences appear so much less likely than they really are. In fact, 1/216 is not the probability of throwing 3 sixes in 6 tries—it's the probability of throwing 3 sixes in 3 tries! Most people correctly intuit that one is more likely to roll 3 sixes in 6 tries than to roll 3 sixes in 3 tries, and indeed the probability of the former is equal to 20/216, or a bit less than 1/10.

And what if your friend had thrown not 3, but 4, 5, or even 6 sixes? The probability one should compute to see if cheating has occurred is the sum of the probabilities for each of these increasingly unlikely possibilities. As it turns out, the probabilities for 4, 5, or 6 sixes occurring in 6 rolls of a die are tiny, but they make a difference: the probability of throwing 3 or more sixes in 6 tries is not slightly under, but slightly over 1/10. In other words, such a result can be expected to occur reasonably often.

The incriminating witness in one of the most famous trials of the nineteenth century made exactly the error described above, and it took ten years and a team of the greatest mathematicians of the time to convince the world that it was wrong.

The Dreyfus Affair: Spy or Scapegoat?

December 22, 1894: Alfred Dreyfus, a captain in the French army, is tried by court-martial and declared guilty of high treason by the unanimous verdict of a jury of seven officers, who sentence him to life imprisonment in perpetual isolation on Devil's Island, a disease- and mosquito-infested rock off the coast of French Guiana.

The incriminating evidence: the "bordereau," an unsigned letter found in the wastebasket of Maximilian von Schwartzkoppen, military attaché at the German Embassy, offering to sell him certain French military documents given in the form of a numbered list.

The main witnesses for the prosecution: Major Armand du Paty de Clam of the army's secret service, who, charged with obtaining a sample of Dreyfus' writing to compare to the bordereau, tells the jury how Dreyfus' fingers trembled as he came into the room, revealing a guilty conscience,* while at the

---

*Later, Gribelin, one of the officers present at the scene, would state that Dreyfus claimed that he was shivering because it was extremely cold outside, but that this claim must have been false, as it had been a fine October day, and furthermore there was a good fire in the office. Joseph Reinach, one of Dreyfus' staunchest defenders and the author of a grand *History of the Dreyfus Affair*, responded to Gribelin's contention with the famous line of repartee: "It is good to avoid too much detail when lying. If it was so fine outside, why did you have to have a good fire? Are you in the habit of lighting fires in summer?" after which he brought out a newspaper dating back to that very October day and showed that the early morning temperature was a chilly 5 degrees Centigrade (41 degrees Fahrenheit).

same time his calm demeanor betrayed the emotional mastery of a practiced deceiver. Du Paty's colleague Major Hubert-Joseph Henry, who recounts how the discovery of the bordereau exposed the existence of a spy within the army, and then, at a loss for further proof, flings his arm dramatically in the direction of the accused, bellowing "And the traitor is—*here!*" And finally, five handwriting experts, two of whom claim that the bordereau could not have been written by Dreyfus, while the other three claim that it was.

The secret: a file transmitted illegally to the jury of officers during their deliberations, a file containing papers apparently so sensitive that for reasons of state security they could not be shown to the accused or to his lawyer. One of the documents—a letter from an Italian military attaché, Alessandro Panizzardi, who was involved in an intense homosexual affair with the German Schwartzkoppen (addressing him as "Maximilienne" and signing his letters "your Alexandrine")—contained a reference to someone he called "that scoundrel D." The others were months-old reports from police informers, carefully modified just before the court-martial to add intimations that the police knew about a French spy delivering material to the Germans.

The facts no one mentioned at the trial: Dreyfus was an Alsatian at a time of French suspicion and resentment of Germany, which had annexed the region of Alsace in 1871. Moreover, he was Jewish at a time of rampant anti-Semitism. It also happened that Dreyfus was an intensely patriotic man, imbued with a deep sense of personal honor. But under the circumstances, that characteristic was ignored, or worse, taken as an insincere pose. After all, Dreyfus was the ideal outsider, and therefore the ideal spy. Or the ideal scapegoat.

ALFRED DREYFUS was subjected to a public degradation ceremony that has lingered in France's collective memory as among the nation's most shameful moments. While Dreyfus repeatedly protested his innocence and proclaimed his love of his country, an adjutant of the Republican Guard broke his sword, tore the epaulettes and buttons from his uniform, and flung them to the ground. All of this was accompanied by screams of "Kill the traitor!" and "Hang the Jew!" emanating from the crowd pressed up against the wrought-iron grilles enclosing the courtyard. Soon after this, the disgraced captain was transferred onto a boat and locked in a barred cell on deck, open to the freezing February winds of the Atlantic. It took the boat two weeks to arrive at Devil's Island.

During the four years that Dreyfus spent on that tiny, sweltering, fever-infested rock, he was the only prisoner there, living in a bare hut under twenty-four-hour surveillance by guards with whom he was not allowed to exchange a single word; deprived of even the basic utensils with which to cook and wash; forbidden to approach the sea (let alone bathe in it); and kept in total ignorance of the efforts being made on his behalf by his wife, Lucie, his brother Mathieu, and his friends. As events in France escalated, Dreyfus was treated with increasing harshness. False reports of an escape attempt that were circulated in French newspapers led to his being shackled to his bed at night by iron rings, and to the construction of an enormous fence around his hut that cut off his view of the sea and restricted his movements to a small rectangle devoid of vegetation. Although Lucie was allowed to write to him and he to her, their letters were censored and delivered months after they were sent. When he was ill, which was most of the time, Dreyfus was allowed to be examined by a doctor from the mainland, but was then forbidden to follow any of the doctor's recommendations concerning diet, hygiene, and sea bathing. His brother attempted to improve his diet by arranging for a shopkeeper in Cayenne to send over canned goods, but the shopkeeper was so harassed by police that he gave up the project.

Except during moments of particular courage, Alfred Dreyfus believed that he was going to die on Devil's Island. Forbidden to receive any news, and ignorant of the enormous effect his case was having on French politics and society, he had to content himself with writing letter after letter to everyone—from his wife, Lucie, up through the chief of the general staff to the president of the Republic, protesting his innocence and begging that the true traitor continue to be hunted. Apart from Lucie, he received no replies.

Actually, the true traitor was being hunted—not by the government, or by the army, but by Lucie and Mathieu Dreyfus with the help of the few public figures whom they had managed to convert to their cause. But even as the net drew closer around him, Major Charles-Ferdinand Walsin-Ester-hazy, the true author of the bordereau, didn't turn a hair. A freewheeling and pleasure-loving scamp, he didn't let the uproar around Dreyfus bother him, but continued lightheartedly with his usual behavior, which consisted of any activity that might earn him a few francs, including gambling, cheating, begging, and informing—as well as spying—along with the frequent seduction of women, particularly the mistresses of his friends.

If the arrest and condemnation of Dreyfus had stopped Esterhazy in his tracks, it is possible that the truth might never have emerged. But Esterhazy didn't know that the document that had convicted Dreyfus was the bordereau written by his own hand; the document had been kept secret so that only Dreyfus, his lawyer, and the judges and witnesses at the court-martial had been allowed to see it. So Esterhazy kept up his amateurish spying activities, which brought such insignificant documents to the German Embassy that finally Schwartzkoppen had enough of them. The German officer penned a telegram (handwritten, in those days, on special thin blue paper) informing his correspondent in thinly veiled language that without some more useful information, their relations might not be worth continuing.

He addressed the telegram (*le petit bleu*, as the little blue telegraph papers were called) to Major Esterhazy by name, but then, perhaps regretting the severity of his language, he ripped it to shreds, dropped it in his wastebasket, and sent off another, gentler version. The shreds were gathered up by the cleaning lady, who delivered them as usual straight to the offices of the French army's secret service.

The head of the secret service who had presided over the Dreyfus affair had recently retired. Major Henry was eager to replace him but had been passed over in favor of Colonel Georges Picquart, an Alsatian known for his rigor and honesty. The army would bitterly regret this decision later, as Picquart, or Picquart's probity, turned out to be an unexpected thorn in their side.

Picquart had been one of the two government observers allowed to remain in the room when Dreyfus' trial was declared closed to the public. He had presided at the ceremony during which the condemned traitor's military insignia were publicly torn off and broken. He had not questioned Dreyfus' guilt for a single moment. Nor did he do so even when the little blue telegram, its thirty or forty shreds attached together with thin bands of transparent paper, was brought to his attention. His first thought was that there must be a second spy at work. But unlike the previous case, here he had a name to go on, and nothing more to do than to keep an eye on Major Esterhazy in order to catch him in the act. In an effort to determine the extent of his spying activities, Picquart had Esterhazy followed, his flat searched, and his mail intercepted. Thus, a couple of letters written by Esterhazy ended up in Picquart's hands. And when he saw them he received a tremendous shock. He recognized the handwriting. It was exactly that of the bordereau. Not similar, or somehow related, as Dreyfus' handwriting had seemed. He set the two documents side by side, Esterhazy's letter and the bordereau, and stared at them.

"I was horrified," he later wrote. "They were not just similar. They were identical."

In the months since Picquart had become head of the secret service, he had never once asked to look at the file that had been secretly communicated to the judges at the Dreyfus trial during their deliberation and that, as rumor had it, had tipped the balance of the verdict to guilt. But he asked for it now and looked inside. He saw one letter referring to someone as "that scoundrel D." He saw some police reports that obviously had been modified to make vague reference to a spy somewhere. That was all he saw. There was nothing else to see.

Picquart summoned handwriting expert Alphonse Bertillon, head of the judicial identity department of the police, who had testified with the utmost certainty that the handwriting of the bordereau was that of Dreyfus, giving what he called an absolutely mathematical proof. Picquart showed Bertillon the bordereau and Esterhazy's letter, with the signature hidden, and asked him what he thought.

"Those were written by the same person," said Bertillon at once.

"But this is a letter that was written very recently," Picquart informed him.

"Well, then," came the immediate response, "the Jews must have been training someone to write exactly like this for a whole year!"

ALTHOUGH PREJUDICED and perhaps a little mad, Alphonse Bertillon was no idiot. He founded the first police laboratory in France for criminal identification and invented some remarkable methods for identifying repeat criminals. Indeed, it was particularly easy in the nineteenth century for a convicted murderer to leave prison, change his identity, and strike again, and extremely difficult to prove that he was the same person as the previous convict. Until the early part of the nineteenth century, physical

branding of prisoners had served as a method for identifying repeat offenders, but its abolishment in 1832 left the police with a serious problem.
In response, Bertillon invented the ingenious method of "anthropometry,"
which consisted in taking fourteen precise measurements of the body of
every convicted criminal, measurements that, he claimed, could identify
any person in a population of 300 million. By writing down the measurements of every criminal in his or her police file, he was able to identify
certain repeat offenders, such as the elusive "Ravachol," an anarchist responsible for innumerable deaths by bomb explosions in and around Paris.
A few years later, in 1902, Bertillon was the first police investigator ever
to identify a murderer by the fingerprints that he had left on a smashed
windowpane.

Bertillon was not a specialist in handwriting analysis, but when the army
invited him to contribute his expertise to identifying the author of the bordereau, he flung himself with zeal into the task. To prove that it could be no
other than Dreyfus, he built up an extraordinary, well-argued theory that
Dreyfus had purposely *forged an imitation of his own handwriting* so that if
he were caught, he could attempt to explain away any evidence against him
by claiming he had been framed. Bertillon decided that the method used
by Dreyfus to imitate yet disguise his own handwriting consisted of tracing
over certain of his own words or parts of words from other letters and documents, and certain words written by other members of his family. He also
asserted that there was far more to the bordereau than the rather anodyne
information it contained. Certain unexplained dots on the page were separated from each other, he claimed, by a distance that was exactly one one-
hundred-thousandth part of a distance found on important and secret
military maps. Then there were some apparent pinpricks or irregularities in
the paper, which he interpreted as the places where Dreyfus had pinned the
thin onionskin paper over the other documents to trace their words. Various
other features of the bordereau indicated a definite method or code, intended to transmit a great deal more information than the written words. In
pursuing this direction, Bertillon applied his own personal version of probability theory to reliable studies made by real experts in the field of military
cryptography and code breaking.

From these studies he gleaned two important pieces of information. One
was that coded messages are sometimes written using a "key," which at that

time often consisted of a single secret word written again and again. The other was that the seven letters that appear most frequently in French writing are *e, n, a, i, r, s,* and *t.*

Having noticed that the words *intéressants* and *intéresse* in the bordereau appeared in handwriting similar to a copy of the word *intérêt* that he found written in a letter on Dreyfus' desk at home, and that this particular word contains five of the seven most common letters of the French alphabet, he made some tests on the possibility of the word *intérêt* being the key on which the whole bordereau was built.

He traced the word *intérêt* from the letter from the desk repeatedly along the lines of a piece of white paper, with no spaces between the words:

Intérêtintérêtintérêtintérêtintérêtintérêtintérêtintérêtintérêtintérêtintérêt

Then he placed the semitransparent bordereau over it. To his amazement, he saw that many of the letters in the bordereau exactly overlaid the letters in his "key," although of course most did not. He noticed that if he shifted the bordereau over by a little more than one millimeter, the letters that fit previously no longer did so, but a large number of those that hadn't fit now did. So he made two keys, one in red and one in green, which were identical except for the thin vertical lines he drew on them, the same distance apart as those of the bordereau.

When he placed the bordereau over the red key with the vertical lines matching up, many letters were overlaid and many others were not properly placed. When he laid it over the green key, the same thing occurred, but with different letters, of course. The two keys were identical except that all the vertical lines were very slightly displaced.

Bertillon then counted the letters *e, n, r, t* of the bordereau that overlaid the exact same letters in the word *intérêt* on the two keys underneath. He was astounded and delighted to discover what he considered an enormous number of these "coincidences," far beyond, as he said, what the probability calculation would justify. In his report to the court, he wrote:

Instead of being found to lie on top of the *t* of *intérêt* 7 times, it is found 15 times; on the first *é*, instead of 26, one finds 40, on the *r*, 20 instead of 9, on the second *ê*, 39 instead of 19, on the final *t*, 10 instead of 6.

The *n* is an exception since instead of finding 11, one finds only 10, but in reality, they are placed over the *r*, since *n* is almost always preceded and followed by a vowel which is placed over the *e*. Indeed, on top of the *r*, instead of finding 8, we find 17.

Such methods may appear of dubious utility, but in Bertillon's mind, the high probability figures gave rise to a strong suspicion of tracing.

The expected probabilities of 7, 26, 9, 19, 6, and so forth were calculated according to the frequency of occurrence of those letters in the key. Consider, for example, the letter *t*. The bordereau contains about 800 letters, and 49 of them are *t*'s. Since two of the letters in the word *intérêt* are *t*'s, one would expect to find 2/7 of the 49 *t*'s in the bordereau lying over *t*'s in the key, making about 14. Of these, one would expect about half of them, 7, to lie over the first *t* in *intérêt*. Instead, Bertillon found 15.

Similarly, there are about 60 *r*'s in the bordereau and only one in the word *intérêt*, so one would expect to find 1/7 of those *r*'s from the bordereau lying over an *r* in the key, which is 8 or 9 *r*'s. Instead, Bertillon found 17. The French analyst was wildly excited by this discovery and became absolutely convinced that his key had been used to write the message.

The fallacy of Bertillon's reasoning is incredibly simple, yet it was not revealed until 1904 by three famous French mathematicians: Henri Poincaré, Paul Appell, and Gaston Darboux. They pointed out that when Bertillon used a single key, say the red one, he found about the expected number of correspondences (same letter over same letter). But he did not want to count only these, because too many of the letters in the bordereau were not even superimposed over letters of the key at all, let alone the correct letters. The only way to get all the letters of the bordereau to be superimposed over letters of the key was the way he did it: two identical keys, one red, one green, slightly shifted with respect to each other. What Bertillon failed to realize is that *by using two keys, he was doubling the probability of coincidence of certain letters lying over the same letters.*

Think about it for a moment. If you take the two copies of the key and shift one of them over by about one letter, and then you count every time an *r* of the bordereau appears over an *r* of the red key, or an *r* of the green key—the *r*'s of those keys *never* being on top of each other, because you've shifted the whole key over by about one place—you are simply counting double the number of *r*'s.

i	n	t	é	r	ê	t	i	n	t	é	r	ê	t	i	n	t	é	r	ê	t	i	n	t	é	r	ê	t
J	e		v	a	i	s		p	a	r	t	i	r		e	n		m	a	n	o	e	u	v	r	e	s
i	n	t	é	r	ê	t	i	n	t	é	r	ê	t	i	n	t	é	r	ê	t	i	n	t	é	r	ê	t	i

Just take the example above, looking at the last line of the bordereau, *Je vais partir en manoeuvres.*

This sentence contains three *r*'s. Using the top key, the *r* of "manoeuvres" corresponds to an *r* of the key. Using the bottom key, it no longer corresponds, of course, but now the first *r* of "partir" corresponds to an *r* of the key.

Using a single key, one *r* out of three corresponds to the key, which is close enough to the expected one out of seven, given that there are only three *r*'s in the sentence anyway. But counting both keys, we find that instead of the expected one *r* out of seven, we get two-thirds (two out of three) *r*'s corresponding. This is obviously much higher than expected. If this happened for every letter, we should start feeling suspicious—until we realize that it simply happened because we're using two keys, not one, and thus doubling the probability.

BERTILLON'S TESTIMONY contributed to Dreyfus' conviction. It was years before the details of Bertillon's analysis were examined closely. In the meantime, Dreyfus' family and supporters continued to fight against all odds for recognition of his innocence. Thanks to their efforts, the story eventually gained such a high profile that it reached the dimensions of a national crisis. One of the most important architects of that crisis was Colonel Picquart, who realized—on sight, with no need for either expertise or calculation—that the handwriting on the bordereau belonged to Esterhazy, not to Dreyfus.

Less convinced by Bertillon's explanations than by the evidence in front of his eyes, Picquart went to his army superiors, Generals Charles Arthur Gonse and Raoul Le Mouton de Boisdeffre, and to the minister of war. He was anxious to share with them the two pieces of news he had discovered: the existence of an active spy whose identity was actually known, and the innocence of Alfred Dreyfus.

To Picquart's shock, he discovered that his superiors had no desire to acknowledge that the army had been guilty of a miscarriage of justice. Instead of receiving praise and honor, instead of securing the real spy's immediate arrest, Picquart found himself dispatched on a long trip to the south of France, with multiple duties to keep him busy, and then on to Tunisia for an indefinite length of time. Major Henry took over Picquart's duties during his absence, and it was made quite clear to the major that his role was to calm what could potentially grow into a large storm.

Only too willing to help, Henry took home some of the papers that had been found in Schwartzkoppen's wastebasket, and with the help of his wife, did a little "cooking"—that is to say, a little cutting, a bit of forging, a dash of imitating, and a dose of tearing off bits of paper and reattaching them differently. What resulted was a "letter" to Schwartzkoppen whose beginning and whose signature, "Alexandrine," were legitimate—ripped from a real letter to Schwartzkoppen from the Italian spy Panizzardi—but whose body contained a sentence straight out of Henry's head, about how they must never ever admit to having had any dealings with the Jew Dreyfus. Henry took care to write the name *Dreyfus* out in full. After all, it didn't appear in a single other document involved in the case.

Henry showed this letter to Generals Gonse and Boisdeffre, his superiors and Picquart's, and they were pleased. History does not recount whether or not they fully realized what the major had been up to, but in any case they probably avoided even asking themselves the question. They made a careful copy of the letter by hand, signed it with their names and ranks, and added it to the file for anyone who might care to have a look. The original fake was put in a different file, a secret one.

All this time, Dreyfus was suffering on Devil's Island, writing letters and struggling with severe illness and depression; Picquart was traveling from place to place and worrying about what he should do; Esterhazy was gambling, cheating, stealing, and spying; and Lucie and Mathieu Dreyfus were knocking on doors trying, with little success, to drum up sympathy for their cause. But no one cared. No one was thinking about Dreyfus yet; the country had other concerns.

Then a friend suggested leaking news of an escape attempt to the newspapers in an attempt to bring the story back to the front pages. The maneuver worked to gain publicity, but it also backfired badly, as a spate of virulent anti-Semitic articles followed, and the aforementioned punishments were inflicted on Dreyfus on his lonely island. The idea would have been a catastrophe but for one important consequence: in the heat of the moment, one person saw a way to make a little money. One of the handwriting experts who had testified at the trial still possessed the photograph of the bordereau that he had worked on, and he sold it to a newspaper, which printed it.

For the first time, Lucie Dreyfus saw with her own eyes the document that had convicted her husband. She saw that the handwriting, although similar, was not her husband's, and she saw black-and-white proof of the terrible injustice that had been committed against him. She and Mathieu hatched a simple but effective plan. They had the bordereau printed on thousands of flyers, together with a sample of Dreyfus' own handwriting and a declaration of his innocence, and distributed them all over Paris, where they were sold at newspapers stands in the manner of the British broadsheets of the nineteenth century.

Their tactic bore fruit. In November 1897, a stockbroker happened to buy the flyer and immediately recognized the handwriting of one of his clients. He contacted Mathieu Dreyfus and showed him correspondence from Major Esterhazy. Mathieu sat down and studied it letter by letter, comparing the dotting of the *i*'s, the turning of the *s*'s, the crossing of the *t*'s with those of the bordereau. When he got up, he, too, knew the name of the real traitor.

BEFORE THIS discovery, the truth about Esterhazy had been known only to a select group of people. There were those who were informed by Colonel Picquart, namely Generals Gonse and Boisdeffre and General Jean-Baptiste Billot, the minister of war, and the two devoted underlings, Majors du Paty de Clam and Henry, who had been instrumental in Dreyfus' condemnation. Apart from these, only two others were aware of the situation: a close friend of Picquart's, the lawyer Louis Leblois, and another important member of the government, who was informed by Leblois reluctantly but as a matter of conscience when he came to realize that Picquart was being purposely maintained at a distance. This was the situation when the name of the traitor was discovered by Mathieu and Lucie Dreyfus, and it was at this point that Gonse, Billot, du Paty, and Henry made the decision to "save the country," which meant going to any length necessary to prevent the judicial error against Dreyfus from coming to light. Thus, it was decided among the four military officers that it was necessary to fatten up the secret file containing proof against Dreyfus.

This task was again left to Major Henry, who—willingly, if rather crudely and incompetently—modified dates and names in a number of compromising documents taken from Schwartzkoppen's wastebasket to make them appear older than they really were. At the same time, Henry amused himself by writing anonymous spy-style letters about "secret combinations," "hidden documents," and "the Syndicate" (by which term he referred to a supposed Jewish lobby seeking to free Dreyfus and destroy the army) and addressing them to Dreyfus on Devil's Island. He then arranged for them to be intercepted and "discovered" en route. And in case Colonel Picquart was intending to return to Paris and tell his side of the story, Henry addressed a few telegrams in a similar style to him, so as to be able to accuse Picquart of engaging in spying activities himself if it should become necessary to discredit him. Unfortunately for Picquart, he had a married mistress at the time who also wrote him secret letters, which were mailed by a mutual friend and contained a similar type of veiled reference. Having taken to intercepting all of Picquart's correspondence, Henry soon discovered this affair, and it made him all the happier. But even this was not enough.

The conspirators decided that Esterhazy must be protected at all costs, for if his guilt were known, then the truth about Dreyfus' innocence—and the army's error—would follow, causing irreparable harm to the country's integrity and prestige. And so began a process unheard of in the annals of political history: members of the government began aiding and covering up for a known spy. The indefatigable Majors du Paty de Clam and Henry took to arranging meetings with Esterhazy in secret nooks and shady corners. They encouraged him to write to the president of the Republic declaring his innocence. They even gave him papers from the Dreyfus file and helped him write letters to the government in which he declared that Picquart was leaking or selling these top-secret papers, and virtually demanded ransom in protective measures in order to return them.

Under normal circumstances the aggressive blackmail that Esterhazy addressed to Félix Faure, president of the Republic, would have landed him in jail. Instead, he was surrounded by a halo of honor. Esterhazy and his helpers also spread rumors that he was being framed as a substitute for the traitor Dreyfus by the Jewish lobby, but that he was being aided in secret by a veiled lady who delivered valuable papers to him out of undeclared love. Newspapers published letters defending the base attacks by the so-called

Syndicate against the honorable soldier Esterhazy. Meanwhile, press friendly to Mathieu Dreyfus printed articles calling for a revision of the Dreyfus trial, a reexamination of the handwriting on the bordereau, and a public rendering of the true circumstances of his conviction.

Slowly but surely the situation began to heat up as people started questioning the prevailing story. There were rumblings in Parliament. Famous journalists and writers began to shift their sympathies. Public opinion became starkly divided between those who were horrified by the jingoistic nationalism and blind respect for authority that were leading the government into denial of an obvious error and those who saw the case as an example of how the Jewish element was undermining and rotting the country from within, trying to save one of their own by smearing a good and honest soldier.

Esterhazy decided to seize the bull by the horns and demanded a court-martial to prove his innocence. The generals happily accepted, hoping to put the matter to rest once and for all. However, all did not go as smoothly as they had hoped. During the lead-up to the trial, the press made public numerous pieces of information that could have potentially destroyed their case.

First to appear was the letter that Major Henry had forged on the part of the Italian military attaché Alessandro Panizzardi. Panizzardi publicly denied writing it and demanded to be heard as a witness at the trial. This caused a moment of agitation for the conspirators, but the army found a solution in refusing to hear him on the grounds that as a known spy, nothing Panizzardi said could be believed.

But then the newspapers began expressing passionate interest in the testimony of Colonel Picquart, who had been recalled to Paris as a witness. This seriously worried the generals, but they decreed that his testimony would be "dangerous to public safety" and must therefore be held behind closed doors. Once again, danger was averted and all was under control.

But now, to the horror of Esterhazy and his supporters, a former mistress of Esterhazy's, who had turned against him because he had stolen and dilapidated most of her modest fortune, gave the newspapers a series of letters he had written to her fifteen years earlier, expressing contempt and hatred for the French army and the French people! "Paris, stormed, defeated and delivered to the looting and pillage of a hundred thousand drunken soldiers—that is a celebration I dream of!" he had written in some kind of wild delirium of anger. "I would never hurt a little dog, but I would

have a hundred thousand French people killed with pleasure!" This letter—
the worst of a bad lot—was published in the daily newspaper *Le Figaro* the
very next day.

Esterhazy panicked and initially claimed it was a forgery. But when his
former mistress revealed that she had a great many other similar ones, he
backpedaled, admitting that he had written the letters, but insisting that the
above passages had been added by a forger's hand. The anti-Dreyfus journalists were embarrassed and fished desperately for excuses. One described
the letters as the product of a "bitter, exalted nature in an access of anger."
Another claimed that the letters had no relation with the spying affair anyway and should be ignored. A third criticized the "nasty woman who sold
for money the letters that her officer friend had written to her in trusting
confidence."

The generals hastily hired some friendly handwriting experts and waited
tensely for their conclusions, which soon emerged as hoped: the looting letter was declared to be merely an imitation of Esterhazy's handwriting. There
was no need to openly name the Jewish "Syndicate" as the instigator of these
forgeries. Everyone knew what everyone else was thinking.

Esterhazy's court-martial opened to a packed hall. Esterhazy was interrogated and responded to all questions with his head held high in the role
of the noble and unjustly slandered officer. Picquart testified behind closed
doors; he told his story, but only the generals heard it. Majors Henry and du
Paty de Clam and their assistants swore that they had seen Picquart in his
office, fabricating forged papers and discussing the secret file with his friend
Leblois, and that Picquart had asked them to lie. Esterhazy's lawyer gave an
impassioned five-hour speech in his defense.

After a three-minute deliberation, Esterhazy was acquitted and carried
in triumph back to the jail, from which he was released in an organized ceremony. Hundreds of spectators lined the street. A powerful bass voice was
heard to shout, "Hats off to the martyr of the Jews!" Heads were uncovered
up and down the street, all the way to the back of the crowd. The next day,
Colonel Georges Picquart was arrested and sent to prison.

THE ARMY had won yet again. But this time their triumph appeared so egre-
gious that it raised the indignation of a larger group of protesters than before.
Most noticeably, the ranks of the Dreyfus supporters were joined by the

highly successful author Émile Zola, who suddenly leaped into the fray with a panache that no one associated with the case had yet been able to muster.

J'ACCUSE! . . .

The headline blazoned across the newspaper the day after Esterhazy's acquittal. And the article that followed, actually an open letter to the president of the Republic, contained an astoundingly accurate description of the mechanisms at work behind the twists and turns of the case, written in a language the incisive and inflammatory power of

Émile Zola, the accuser

which could only have been produced by a truly great writer.

Zola's "J'accuse" remains one of the seminal texts of French literature. On the hundredth anniversary of its publication, a copy of the text two stories high was hung on the National Assembly building in Paris. Entering the words "j'accuse" into Google will bring hundreds of thousands of hits leading directly to Zola's unforgettable words. After describing his view of the case and the danger of dishonor that hung over the country, Zola ended the article by pointing his finger directly at those he considered the guiltiest.

> I accuse Lt. Col. du Paty de Clam of being the diabolical creator of this miscarriage of justice—unwittingly, I would like to believe—and of defending this sorry deed, over the last three years, by all manner of ludicrous and evil machinations.
>
> I accuse General Mercier of complicity, at least by mental weakness, in one of the greatest iniquities of the century.
>
> I accuse General Billot of having held in his hands absolute proof of Dreyfus's innocence and covering it up, and making himself guilty of this crime against mankind and justice, as a political expedient and a way for the compromised General Staff to save face.

He goes on to address others complicit in the crime, including Generals Boisdeffre and Gonse, the handwriting experts who examined Esterhazy's letter, the War Office, the first court-martial for convicting Dreyfus on secret information, and the second one for knowingly acquitting a guilty man. His project is not mere finger-pointing, but rather the search for truth:

> As for the people I am accusing, I do not know them, I have never seen them, and I bear them neither ill will nor hatred. To me they are mere entities, agents of harm to society. The action I am taking is no more than a radical measure to hasten the explosion of truth and justice.
>
> I have but one passion: to enlighten those who have been kept in the dark, in the name of humanity which has suffered so much and is entitled to happiness. My fiery protest is simply the cry of my very soul. Let them dare, then, to bring me before a court of law and let the enquiry take place in broad daylight! I am waiting.

As Zola had anticipated and indeed desired, the War Office sued him for slander, and in February 1898 he was sent to trial. Together with his lawyers and the increasingly large group of influential intellectuals supporting the cause of innocence, he worked to turn his own trial into a trial of the army. Two hundred witnesses were summoned; every single person associated with the case in any way was called to the stand. The country's best experts from schools devoted to the study of manuscripts testified that the handwriting of the bordereau was Esterhazy's. Alphonse Bertillon alone testified to the contrary, repeating his "geometric" theory, but it was so complicated that some members of the public were actually heard laughing incredulously.

Colonel Picquart was brought from his prison cell for cross-examination and told the court the story of the discovery of the *petit bleu,* of how he had realized first that Esterhazy was a spy, and then that he was the author of the bordereau. He also related how he had informed his superiors of his findings only to be sent away, and had subsequently become the victim of a campaign of lies and harassment ending with his arrest.

Major du Paty de Clam—now promoted to lieutenant colonel—testified, Major Henry testified, the generals concerned in the case were questioned, and Esterhazy came to testify. Hundreds of questions were blocked by the judge, who intervened before the witnesses had time to respond with the oft-repeated words: "The question shall not be asked." For the questions that were allowed, the witnesses all invoked the necessity to remain silent in order to protect the country's safety. Esterhazy himself, after filling the newspapers with inflammatory exclamations about "something" he was going to do that would "fill the streets of Paris with corpses!" was compelled to remain silent in obedience to his protectors in the army. He listened without response to the increasingly pointed questions about whether he was the author of the letter to his former lover, where he obtained his money, whether he had known Schwartzkoppen, whether he had written the bordereau, and whether he had ever been paid for spying, all while gripping the bar with white-knuckled hands. The generals avoided answering all direct questions by invoking the higher interest, the honor of the country, national security, and the necessity of preserving the good name of France on the international scene.*

One of the fieriest generals ended the trial by giving a speech evoking the terrifying consequences that a lack of confidence in the army would have on the nation.

> Then what do you expect the army to be, on the day of danger, which may come sooner than you think? What do you want the poor soldiers to do, led into battle by chiefs that have been discredited in their eyes? Our sons will be led into a butchery, members of the Jury! But Mr. Zola here, he will have won his own battle. He will write a book about the defeat—he will transport the French language into every corner of the universe—and France will have been struck off the map on that day!

The public was on its feet. The jury was in tears. It took them just thirty-five minutes to convict Zola.

---

*If France's good name on the international scene was to be preserved by insisting that Dreyfus was guilty, then other countries were not aware of it. During the course of the trial, German, English, American, Italian, Spanish, and Dutch newspapers expressed the most extreme astonishment at the blindness of the French government, which was crushing the notion of justice underfoot, arbitrarily turning it into a concept that opposed, rather than supported, the ideal of nationalism. "The French are hypnotized with fear of the truth," wrote a Russian newspaper. "Mr. Zola's crime was to stand up in defense of truth and civil liberty!" screamed the *Times* of London. "Europe must defend the values of France against France itself," appeared in the Belgian headlines. And so on and so forth.

Condemned on appeal to a year in prison, Zola fled to England, where he was homesick and miserable. The army breathed a sigh of relief. The pro-Dreyfus forces were in disarray, and it seemed unlikely that they could regain their lost terrain in the face of massive public disapproval. Partly due to the Dreyfus affair and the giant wave of anti-Semitism it unleashed, the May 1898 parliamentary elections swept a powerful reactionary, nationalistic, anti-Semitic, and anti-Dreyfus faction to power. The year 1898 saw an incredible polarization of society as people took sides on the issue, fracturing the country down a line that cut across distinctions due to social class, profession, or age. A famous newspaper drawing from 1898 shows two scenes from a family dinner, the first with the caption, "Welcome! Let's not talk about the Dreyfus affair," and the second, "They talked about it."

Around this time General Billot, the minister of war who had overseen every phase of the Dreyfus affair, was replaced by the willful and charismatic Godefroy Cavaignac. Cavaignac was appointed because of his intransigence, his ability to influence people by sheer force of will, and his staunchly anti-Dreyfus position. But General Billot and the military elite who had been involved in the affair did not realize what trouble such a personality could bring.

In the end, it was the same indomitable will that had made Cavaignac so attractive to the government, that desire to have full control, that brought the whole house of cards down. Irritated by the continued public words and actions of the reduced but vocal group of Dreyfus supporters, Cavaignac announced his intention to put a smashing end to the entire affair. His first idea was to organize a massive trial in which every notable Dreyfusard, from Lucie and Mathieu to Zola and Picquart, together with all the lawyers who had worked for them, and all Dreyfus-friendly journalists and newspaper editors, were to be publicly accused of treason and of conspiring against the good of the country.

Horrified, the army convinced Cavaignac that this project, far from settling the question, would open a Pandora's box too dangerous to contemplate. But he was not in on the whirlpool of cheating, lying, and forgery that had swirled ceaselessly around the case. He believed sincerely in Dreyfus' guilt and in the possibility of obtaining totally incontrovertible proof. Foiled in his grandiose plan, he next decided to go through the Dreyfus file, pick

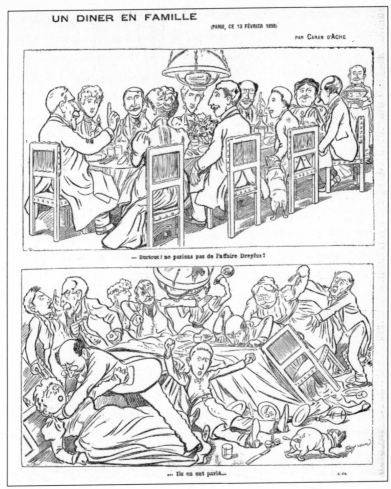

They talked about it

out the most convincing documents, and make them public. He demanded the file and had it examined, classified, and labeled by a young and devoted officer called Captain Cuignet. Cavaignac wanted to see everything, and by now, thanks to Major Henry's indefatigable efforts, there were well over a thousand documents, which filled ten boxes.

He picked out the three documents he considered absolutely damning, and over the weak and slightly desperate objections of Generals Billot, Gonse, and Boisdeffre, he triumphantly carried them into the Chamber of

Deputies and read them aloud from the tribune. One was the original "that scoundrel D." letter from Panizzardi to Schwartzkoppen. The second was a letter from Panizzardi to Schwartzkoppen mentioning a certain "P.," dated from 1896—except that unbeknownst to Cavaignac, Henry had altered the P to a D, and the date to 1894. Finally, the third document selected by Cavaignac was the fake letter from Panizzardi to Schwartzkoppen written by Henry and his wife, containing the name of Dreyfus written out in full.

Not only did Cavaignac read out these documents in public and proclaim his absolute belief in Dreyfus' guilt and the utter impossibility of the case ever being reopened, but also he had his speech, together with a photograph of the last of the three documents, printed on posters that were then placed on the wall of every single town hall in all of France. Thus, he proudly declared, he would stamp with his heel once and for all on the traitorous Jewish snake that kept raising its head. End of story.

Or so he thought. But in fact, as the Dreyfus camp soon realized, he had made some tremendous errors in his rashness. First, breaking away from the army's declaration of absolute respect for "the thing that had been judged," he had as much as admitted that the court-martial's decision could be called into question by an individual. Second, he had shown that Dreyfus had been convicted on documents that were not the one—the single bordereau—that had been introduced against him at his trial. Third, by reading the documents aloud, he had put an end to the notion that they must be kept secret for reasons of national security, and therefore removed any reason to have held or to continue to hold parts of the trials behind closed doors. And finally—this was a major point—he had stated a simple truth, but one that was being denied with increasing vehemence by a large section of both the government and the population: what mattered, after all, was whether or not Dreyfus was guilty, and *not* the fact that questioning his guilt might constitute a blow against the honor of the army and the nation. Cavaignac himself had posed the question of whether or not Dreyfus was guilty. He had answered loudly in the affirmative, to be sure, but the very fact that he had posed the question at all showed that it could be done without a loss of honor to anyone.

The Dreyfus supporters took heart. Cavaignac's actions showed that the time was ripe for Lucie Dreyfus to make an official demand for the revision of her husband's trial. Cavaignac geared himself up to exert all of his influence in order to ensure that her demand be rejected. But he was foiled, for a second time, by an unexpected reaction from the heart of his own side, the army.

During this time, young Captain Cuignet had been continuing his work on the Dreyfus file, examining and classifying the documents one by one with the aim of helping his superiors gain a complete understanding of the case and put a final end to their problems. He worked late into the night. Sitting at his desk, he took up the main folder for the umpteenth time and stared at the papers it contained. The letter naming "that Jew Dreyfus," the one that had been printed on posters and put up in every town hall in the country, particularly attracted his attention. He held it up to the light and turned it this way and that. Indeed, a strong lamplight was necessary to see what Cuignet then saw: the thin writing lines on the beginning and end of the letter were violet in color, while the lines on the paper containing the middle part were blue. The letter was actually composed of two different letters stuck together. Now that Cuignet realized this, he also saw that the handwriting of the middle part was slightly different, as was the pen used. Furthermore, he noticed, Panizzardi had not made quite so many mistakes in his French in any of the other letters he had written.

Cuignet was a firm believer in Dreyfus' guilt, but he was an honest man, and not personally involved in the case. He waited, anxious and unhappy, till the next morning and then took the letter straight to Cavaignac. In his office in the light of day, with the August sun streaming through the windows, Cavaignac couldn't see what Cuignet was talking about. But Cuignet insisted he look more closely. He had the shutters closed, the curtains drawn, and the lamps lit, and he held the letter up close to the light. Then Cavaignac saw the two colors, and he also saw that this was not a moment for him to lie. His credibility was at stake, and he had an honest and independent officer standing in front of him.

"Yes, this letter is a fake," said Cavaignac, but it wasn't an admission of fault; that simply wasn't in his character. It was an attack.

The credit he would gain from public recognition of his truthfulness, he thought, would help his cause rather than hurt it. As for the question of guilt, there were plenty of other incriminating documents in the file—or so he believed. Still, the new discovery that had been made necessitated some kind of exemplary reaction.

Cavaignac called Henry, Gonse, and Boisdeffre into his office, and there he subjected Henry to a severe and pointed interrogation the likes of which he had never experienced before. Major Henry broke down. He admitted that he had done everything to please the two generals, to bring calm to their ruffled spirits. He had done it all for his country. He wept and looked pleadingly at his two erstwhile protectors, but they sat without a word. Silently, Boisdeffre took pen and paper and wrote out a letter of resignation in which he stated that he had been deceived by Major Henry, a man he trusted, and that he felt the major had thereby lost the authority to continue in his position.

Henry was taken immediately to prison, where he drank himself into a stupor, having been provided, surprisingly, with a bottle of rum. He wrote a panicked letter to General Gonse: "Please come to see me. I absolutely have to talk to you." No response. Hours later, Henry wrote a loving letter to his wife, enjoining her to take care of their son and swearing that he had not done anything wrong, but had merely written down facts that he had been told. Leaving this letter on the table, he finished the rum, then picked up another sheet of paper: "My beloved Berthe, I'm going mad, a dreadful pain is squeezing my brain, I'm going to bathe in the Seine . . ." The letter remained unfinished. The officer who came to bring him dinner found Henry lying on his bed, a razor in his hand, his throat slashed, blood flowing over his chest and hands, soaking the sheets, and running in a pool across the floor.

The French supreme court judges accepted Lucie Dreyfus' demand for the revision of her husband's trial, and a boat was sent to Devil's Island to bring Alfred Dreyfus home.

IN ORDER to avoid mass riots, the retrial was held in the city of Rennes, in Normandy, and Dreyfus was kept in prison there while awaiting its start, set for August 7, 1899. Lucie came to Rennes, and husband and wife were allowed to take each other in their arms, to look into each other's faces for the first time in four and a half years. She was shocked to see how changed he was.

Weak, ill, and malnourished, he asked that his military uniform, which he was once again allowed to wear, be thickened with cotton pads in order to give him some semblance of solidity. He forced himself to walk into court with small steps in order to hide his tendency to stagger. Dreyfus wanted to

Alfred Dreyfus at his retrial, 1899

be acquitted on the evidence. He wanted the judicial error of which he had been a victim to be set right. He did not want anybody's pity. Whenever the lawyers tried to discuss the horror and suffering that he had endured on Devil's Island, he cut the debate short in a voice that the journalists who were present found dry and emotionless. He did not make loud or passionate declarations, nor did he want his lawyers to make them. He simply wanted it to be publicly, rationally, and factually proven that he was not the author of the bordereau.

The witnesses were the same who had testified at the original trial in 1894, with the exception, of course, of Major Henry. Some of the handwriting experts were called back to testify again. One in particular, Charavay, made a startling statement that moved and impressed both the jury and the public. "I wish to declare that in 1894, I was misled by a certain resemblance of handwriting into attributing the bordereau to Dreyfus. But since then, having been presented with a new sample of handwriting [Esterhazy's], I have understood my error, and it is a great relief to my conscience to be able to stand here before you, the judges, and above all before him who was the victim of my error, and to declare that I made a mistake in 1894."

But Alphonse Bertillon, he of the devastating "geometric proof" that the bordereau was written by Dreyfus, came back with more of the same. Utterly convinced that Dreyfus had written the bordereau in a purposely modified version of his own handwriting, Bertillon once again explained his theory in detail.

BERTILLON'S THEORY about the construction of the bordereau as self-forgery had never ceased to develop over the five years between 1894 and 1899. As we saw earlier, his "geometric proof" was riddled with serious probabilistic errors. His greatest fallacy was to detect what he called "coincidences" in the bordereau, to miscalculate the probabilities of such "coincidences" arising, and to conclude that they were simply too unlikely not to be the result of a purposeful act.

The onionskin paper the bordereau was written on was lined with nearly invisible, fine vertical fibers spaced exactly half a centimeter apart. Given the width of the stroke made by the pen that was used to write the bordereau, Bertillon calculated that there were about five possible positions that a pen stroke could occupy with respect to these lines: on a line, just touching a line to the left, between two lines but nearer to the left-hand one, between two lines but nearer to the right-hand one, and just touching a line to the right. Thus, he claimed that the probability for a given pen stroke—for example, the initial pen stroke of any given word in the bordereau—to occupy one of these positions was equal to 1/5. So far his deduction is reasonable enough.

Bertillon also chose to restrict his attention to 26 particular words in the bordereau, namely the 13 words of more than one syllable that happened to be repeated more than once within the text of the letter. His explanation was that graphologists concentrate particularly on such words because they afford a greater terrain for comparison. In any case, those words constituted a reasonable sample of all the words in the letter.

He then observed with a magnifying glass the positioning of the initial pen strokes of the initial letters of these 26 words and discovered that out of the 13 pairs of initial pen strokes and the 13 pairs of final pen strokes belonging to the 13 pairs of repeated words, so 26 pairs of pen strokes in all, 8 of these pairs appeared to have a peculiar property: they were all placed identically with respect to the vertical lines running faintly down the paper.

Bertillon then gave the following reasoning, described during his testimony at Dreyfus' second court-martial, and reported in its entirety in *Le Figaro* on August 25, 1899. We include part of that article here, not for the purpose of serious study, but to give a sense of what the readers of *Le Figaro* were subjected to, not to mention the jury members who sat through such argumentation for hours. What nonprofessional has the patience to listen to or read such a speech with enough attention to argue coherently against it?

The striking observation is that when you look at the bordereau, and place on top of it this transparent sheet marked with vertical lines separated by exactly half a centimetre, the repeated words often occur with their initial letters placed exactly the same with respect to these vertical lines.

Take the word *modification* on line 10 and the word *modification* on line 6. The initial pen stroke is at exactly the same distance from the vertical line just to the right of it.

What is the probability that such coincidences could be the fruit of chance? From a practical point of view, how many naturally written letters would you need to have some chance of finding these pairs of words set so similarly? [Murmurs in the court.]

Just consider one of the words I mentioned, at random, say the two occurrences of *modification*.

Once the writer has written this word once, with the *m* just touching one of the invisible vertical lines separated by half-centimeters, what is the probability that he writes the word a second time with the second *m* also just touching one of these lines? Given the thickness of the pen stroke, that probability is about 1/5.

Thus, if we have this bordereau written out ten thousand times naturally, we would only find two thousand copies in which the two letter *m*'s appeared in the same position with respect to the vertical lines.

Now, could this fact imply that the initial *d* of the two occurrences of the word *disposition* should also be in the same position?

Obviously not.

The placing of those two pairs of words is totally independent. Placing one pair can have nothing to do with the placing of the second pair.

Thus, if the first occurrence of the word *disposition* is placed as in the bordereau, then the second one could only be placed in the same way about 1/5 of the time.

We saw that out of 10,000 copies written normally, we had only two thousand with the two occurrences of *modification* placed the same; now we have only 400 in which also the two occurrences of *disposition* are placed the same.

We must re-divide this 400 by 5 because of the two occurrences of the word *manoeuvre*, whose initial *m*'s are also placed identically with respect to the vertical lines. Thus we find only 80 possibilities, which must be again divided by 5, giving 16, because of the two occurrences of the word *copie*. [Murmurs in the court.]

Finally, after dividing again by 5 because the same phenomenon occurs for the word *nouveau*, we find that we have barely 3 chances out of the original 10,000 to have these five coincidences all satisfied at once.

And there are more coincidences that we could give. Thus, we can state that even out of one hundred million copies written naturally, there would hardly be even one or two that could contain all the coincidences shown here.

Conclusion: Whoever the author, whatever the purpose, what we have here is unquestionably an artificially constructed document.

Given the position of the initial (or final) pen stroke of the first occurrence of the repeated word, there is a 1/5 chance that the corresponding pen stroke of the second occurrence of the same word occupies exactly the same position. This coincidence happens 8 times in the bordereau, so the total probability is $(1/5)^8 = .00000256$, which is roughly one chance out of four hundred thousand. That probability is much too small to have been a mere product of chance. Therefore, the placing of so many initial or final letters in equivalent positions must have been done carefully on purpose, and must denote a purposeful intention, probably a secret code.

—*Le Figaro, August 25, 1899*

That little calculation convinced a jury of seven. It is, however, quite a perfect example of the situation with the dice explained in the example of Math Error Number 10 at the beginning of this chapter. First, the probability that Bertillon computed is not that of 8 pairs of letters out of 26 being in identical positions, but 8 pairs out of 8, exactly as in the example of the arrows in the bull's-eye. If he had made the proper computation, he would have come up with a probability of slightly over 7 chances in 100, instead of the 1 in 400,000 he claimed.

On top of that, since Bertillon calculated only the probability of 8 coincidences, forgetting the others, he also necessarily forgot the fact that he would have been equally or even more surprised by 9, 10, 11, or 12 coincidences . . . all the way up to 26. So one should really calculate not the probability of 8 coincidences, but the probability of all numbers equal to or greater than 8, and add them together. Now the result is that Bertillon found surprising and highly suspicious something that actually occurs with a probability of over 13 in 100!

Perhaps 13 in 100 denotes an event that is fairly rare, but certainly not so rare as to make anyone think that such a number of coincidences must be due to a purposeful placement of the letters denoting a secret code. If Bertillon had made the correct mathematical calculation, he would never have made that strange deduction. But at the court-martial of Alfred Dreyfus, there was no one who could see or correct his errors and the specious deductions to which they led him.

WHOM TO believe and whom to doubt? At this point in the trial, the outcome was impossible to predict. One thing was clear, however: an acquittal would lead immediately to accusations of the original creators of the case against Dreyfus: Generals Mercier, Gonse, and Boisdeffre, and Major du Paty de Clam. The price was too great for the army to pay. Emergency measures were necessary, and General Mercier undertook to apply them.

Mercier took the stand and delivered a piece of information that shocked and stunned everyone present, most of all Dreyfus himself. The general claimed that the bordereau analyzed by the experts, written on thin onion-skin paper, was only a traced copy of a secretly kept true original version of the bordereau. This original bordereau, he claimed, was in the possession of the army's secret service, and the absolute, definitive proof of its authenticity was an annotation to that effect by no less a personage than Kaiser Wilhelm II of Germany himself!

Mercier's statement was an outright lie, but no one could prove this, because for the usual "reasons of national security," he refused to produce the document in court. Instead, he gave his word of honor (for what it was worth) that he knew for a fact that Dreyfus was guilty. He stepped down from the stand with the air of a hero saving his country from indignity, dishonor, and defeat at the hands of a despicable enemy.

As a result, the jury brought in a majority verdict of guilty "with extenuating circumstances" for Dreyfus (prompting Dreyfus to exclaim, "What kind of extenuating circumstances exist for treason?!") with two of the judges voting for acquittal. He was condemned to ten years in prison, but spared a repetition of the disgraceful scene of degradation.

The government, aware of the enormous injustice that had been committed, but unable to rectify it without inflicting a public humiliation on the army, entered a phase of intense negotiations with Dreyfus' supporters, which resulted in his being offered a full and immediate pardon, setting him free at once. It was a bargain of sorts: the government would publicly uphold the verdict of the court-martial but explain that it was simply acting out of pity. In return, Dreyfus could return to the life and the family from which he had been so dramatically separated.

It hurt him to accept a pardon, knowing he was innocent, but he was physically unable to endure more imprisonment and desperate to return to his loved ones and help raise his children, who had not seen him for almost five years. Thus Dreyfus accepted the presidential pardon with a letter in which he proclaimed his innocence, and declared his intention of pursuing the full and public revelation of the truth for as long as it would take to obtain it.

It took another seven years of relentless struggle, during which the documents that had been used to condemn him were legally recognized as forgeries, and during which time three of the most famous mathematicians in France—Henri Poincaré, Gaston Darboux, and Paul Appell—wrote a detailed report analyzing each and every one of Bertillon's mathematical errors. Their text ends with the claim that Bertillon's claims are "utterly deprived of scientific value . . . because the rules of probability calculation were not applied correctly."

On July 12, 1906, the French supreme court read aloud a declaration that annulled the Rennes judgment and reinstated Dreyfus in the army. On July 21, 1906, he was awarded the Legion of Honor in a ceremony at the same military school where he had undergone the excruciating degradation in 1894. Eight years later, already fifty-five years old, he fought for his country, participating in some of the worst battles of World War I, including the infamous Battle of Verdun. At the end of the war he retired and devoted the rest of his life to his family and to the history of his appalling affair.

None of the generals and majors involved in knowingly building the case against Dreyfus ever confessed, but their pact of silence could not withstand the march of history, and their role was eventually recognized for what it really was. As for Esterhazy, he went to settle in an English village under an assumed name. His tombstone can still be seen in the churchyard at Harpenden, Hertfordshire, but no one would recognize it as his. It is inscribed:

IN LOVING MEMORY OF

COUNT DE VOILEMENT

1849–1923

HE HAS OUTSOARED THE SHADOW OF OUR NIGHT

CONCLUSION

In nine of the ten cases described in this book, not only did math obscure the truth, but in some cases it led to real miscarriages of justice. The only exception is the Berkeley sex discrimination case, in which the first level of mathematical analysis—examination of the statistics—led to an impression of injustice, but the second level, consisting in a correct breakdown and analysis of the statistics, revealed the true answer. In the other cases, even when the mathematical errors were eventually spotted by experts—Sally Clark, the Collins couple, Joe Sneed, Lucia de Berk, and the Dreyfus affair—the final results were not obtained by correcting the math, but by ignoring it, and sometimes by introducing new evidence; for example, the medical evidence of infection in Sally Clark's son, Sneed's wife's testimony about his violence, the evidence proving that baby Amber did not have an unusual amount of digoxin in her body when she died, and the fact that the document that had originally incriminated Alfred Dreyfus was eventually recognized as having been penned by Esterhazy.

This leads to the question of whether mathematics should be used in trials at all. Should it really have a role in the detection and proof of crime? The obvious disadvantage, which is the major subject of this book, is that it is only too easy for non-mathematicians, or for mathematicians who are not used to applying math in real-life situations, to misunderstand and misuse mathematics in all sorts of different ways.

As a matter of fact, there are a number of people working in law and crime detection who have sought a solution to the problem of the use and role of mathematics in criminal law. Over the last forty years, several scholarly articles have appeared on the subject, published in places like the *Harvard Law Review* and other renowned law journals (see the "Sources" section). However, they are rarely if ever read by mathematicians or by members of the general public.

One of the most famous articles on the subject, perhaps even the best known, is "Trial by Mathematics: Precision and Ritual in the Legal Process," by Laurence Tribe. Tribe is a professor at Harvard Law School and was the young aide who wrote the mathematical portion of the supreme court of California's judgment reversing the Collins conviction. Because of this case and others he encountered over a long career, some of which have been detailed in this book, Tribe has thought deeply about the question of mathematics at trial and has come to the conclusion that the danger of the kind of error we have seen here is too great to make it worth allowing. He fears that the logical and numerical approach used in mathematical thinking is so different from the intuitive approach that must be taken by jury members when evaluating evidence that the two approaches cannot properly be combined. He concludes that mathematics does not belong at trial.

Tribe wrote his article as a beautifully argued, passionate response to an earlier article by Michael Finkelstein and William Fairley, who had proposed a specific example of a situation in which they believed a simple probability calculation could produce an important insight that was unlikely to be grasped properly in any other way. Tribe argues that the situation is rarely if ever so simple that a mathematical model takes all of its human subtleties into account; he raises the double specter of errors made in the mathematics itself and of math that is not wrong but is too simple to apply to the situation at hand. We have seen both types of problem illustrated in this book.

Above and beyond these issues, Tribe fears the psychological effect of mathematics, which can overwhelm jurors, and the use of which "threatens to make the legal system seem even more alien and inhuman than it already does to distressingly many. . . . The need now is to enhance the community comprehension of the trial process, not to exacerbate an already serious problem by shrouding the process in mathematical obscurity." He argues that some of the basic values of our traditional justice system may be lost if mathematics becomes commonly used in the courtroom, and that "guided and perhaps intimidated by the seeming inexorability of numbers, induced by the persuasive force of formulas and the precision of the decimal points to perceive themselves as performing a largely mechanical and automatic role, few jurors could be relied upon to recall, let alone to perform, this humanizing function, to employ their intuition and their sense of community values to shape their ultimate conclusions."

Written forty years ago, Tribe's article has had so strong an influence on courtroom procedures that some claim he single-handedly set back the progress of statistics in the courtroom by decades. During this time it was risky to introduce such arguments in the courtroom and run the danger of having the verdict overturned on appeal with citations of the many cases such as the ones in this book.

Yet in the twenty-first century, probability is making a comeback in the courtroom. The primary reason for this is the omnipresence of DNA analysis, which did not exist when Tribe wrote his article in 1971. In order to understand the deep relationship between DNA analysis and the general use of probability in the courtroom, it is necessary to make an argument that perhaps runs counter to some commonly held views about DNA.

If an identification is made from a high-quality unmixed DNA sample, then it is generally held, inside and outside the courtroom, to be a virtual certainty. Nevertheless, the means used to arrive at such an identification are nothing but the type of probability calculation that we have seen so much of in this book: a statistically established probability for the occurrence in given populations of each of the thirteen genetic loci usually considered in establishing a match, and the product of these (independent) probabilities when several loci are present.

The problems that arise with this type of probability calculation in DNA analysis tend to occur, as we have seen, when the DNA is degraded, partial, or a mixed sample. Then, assuming that the forensic work is done in the most accurate possible manner and without error or carelessness, actual probability calculations in the courtroom become unavoidable, unless the DNA is to be thrown out altogether, which is impossible. And this will increasingly lead to the use of probability theory in other situations as well, on the grounds that it is exactly the same theory; thus there is no reason at all to allow it in one situation and not in another. Thanks to DNA, the mathematics that was ceremoniously chased out of the courtroom by Laurence Tribe and the judges who agreed with him is sneaking in by the back door.

Still, however, as we saw in the cases in this book involving DNA (the murders of Meredith Kercher and Diana Sylvester), the results of any DNA analysis that is not a straightforward, high-quality, single-individual identification are open to question and debate in the courtroom, and the mathematics involved is subject to error in the hands of lawyers. Because math is going to be present in the courtroom as long as forensic analysis takes place, it is becoming a rather urgent problem to establish criteria for its use. At the same time, it is probably going to be necessary to educate the public, from which juries are drawn, to recognize some of the most common mathematical principles that forensic analysis cannot do without. Although Tribe saw the public attitude toward mathematics as a kind of uncomprehending awe, we do not believe that this is really the prevailing attitude toward mathematics. Even if it is, we are convinced that it can be changed incrementally without recourse to any extreme measures. Indeed, the public familiarity with at least some of the basic features of DNA analysis proves that others can become equally familiar, and the frequency and popularity of television series focusing on crime detection prove that people are not indifferent to this theme.

Chief among the probability techniques that are making their appearance more frequently of late is Bayesian reasoning: the use of the so-called Bayes' theorem and its generalization to Bayesian networks. Bayes' theorem has already been used in court many times, but given the lack of a coherent attitude on the part of the courts toward probability, its use has met with varied success. Sometimes it is accepted, other times challenged, and most recently, in a British case dating from July 2011, it was rejected out of hand by the appeal judge at a murder trial, in a judgment that was interpreted by many as rejecting the use of Bayes' theorem at trial in general.

This judgment served as a catalyst for the community of mathematicians and statisticians who are involved with criminal trials, either theoretically or as expert witnesses. An international team led by statisticians at Queen Mary, University of London, the "Bayes and the Law" Research Consortium, founded as a reaction to the anti-Bayes ruling, has begun work on a research project whose goal, drawing on past cases of mathematics in trial, is to put together a set of criteria and a set of analytic tools that should ensure that probability at trial will henceforth be used correctly, applied only to situations in which it can give a meaningful result, and, by virtue of these advantages, be proof against attack at appeal.

It seems that such a plan is the only possible approach to seriously using math at trial without confronting the prejudices, fears, and manipulation that have so often characterized it, as our book rather sadly proves. We are optimistic about the project and hope to follow its progress and share further cases of the use and misuse of math at trial in future books.

SOURCES

A few of the cases studied in this book came to our attention through the media. It was originally the stories of victims Sally Clark, Lucia de Berk, and Meredith Kercher that caught our eye and made us sensitive to the issue of statistics used in trial, as well as the Madoff affair, of course, which led us to Ponzi. The Collins case is evoked in Ben Goldacre's *Bad Science* (Fourth Estate, 2008) and other popular books on the subject. Many thanks are due to Jordan Ellenberg for pointing out the mathematically fascinating case of Diana Sylvester. The importance of mathematics in the Dreyfus affair is known to mathematicians, although not to most historians, let alone the public. The remaining cases—the Berkeley affair, Joe Sneed, and Hetty Green—came up as we deepened our research into the scholarly literature. Many of these cases are cited repeatedly in scholarly articles on the subject of the use of mathematics in trial; the Sneed case, in particular, is a ubiquitous reference. Below is a list of those scholarly references that were the most useful and relevant to us in our research.

For anyone interested in the theoretical aspects of math at trial, the best starting point is the profound and sensitive article by Laurence Tribe, "Trial by Mathematics: Precision and Ritual in the Legal Process" (84 *Harvard Law Review* 1329 [1970–1971]). This article arose as a response to M. Finkelstein and W. Fairley's article "A Bayesian Approach to Identification Evidence" (83 *Harvard Law Review* 489 [1970]). They wrote a rebuttal, "The Continuing Debate over Mathematics in the Law of Evidence" (84 *Harvard Law Review* 1801 [1970–1971]), and Tribe again responded with "A Further Critique of Mathematical Proof" (84 *Harvard Law Review* 1810 [1970–1971]). This fascinating debate among intellectuals of the legal profession provided us with ideas, information, and great stimulation. We found out only later that Tribe studied mathematics as an undergraduate and was directly, although anonymously, involved in the Collins case.

Two other authors have written articles that were both fascinating and helpful to us. David Kaye is a specialist in the subject of math at trial. Apart from an interesting account of the Dreyfus affair, he is also the author of "The Admissibility of 'Probability Evidence' in Criminal Trials," parts 1 and 2 (*Jurimetrics* 26, no. 4 [1986], and 27, no. 2 [1987]), and dozens of other relevant articles. Alan Cullison's "Identification by Probabilities and Trial by Arithmetic: A Lesson for Beginners in How to Be Wrong with Greater Precision" (6 *Houston Law Review* 473 [1968–1969]) also shed light on the many problems with mathematics in trials.

Finally, the book *Applying Statistics in the Courtroom,* by Phillip I. Good (Chapman and Hall, 2001), led us much further into the realm of complicated statistics than we ever meant to venture.

What follows is a list of the documents and references we studied for each individual case.

Chapter 1: The Case of Sally Clark

Biographical material on the life and career of Roy Meadow is available from a large number of online sources, starting with Wikipedia. The website msbp.com (by MAMA, or Mothers Against Munchausen Allegations) contains information from an interview of Roy Meadow during the time when Sally Clark was already in prison. (He says, "I probably have more sympathy for her than the rest of the population.") On his being struck off by the GMC there is again ample information; see, for example, the BBC online news article "Sir Roy Meadow Struck Off by GMC" from July 15, 2005. There is the transcript of Dr. Robert Kaplan's lecture "The Rise and Fall of Sir Roy Meadow." And of course there are Meadow's scholarly publications in medical journals, particularly the seminal "Munchausen Syndrome by Proxy: the hinterland of child abuse," in *Lancet* (August 13, 1977), and the subsequent "What is, and what is not, Munchausen's Syndrome by Proxy," in *Archives of Disease in Childhood,* a medical journal that also contains articles suspicious of MSbP (cf. the articles "Practical concerns about the diagnosis of Munchausen syndrome by proxy" by C. J. Morley, and "Is Munchausen syndrome by proxy really a syndrome?" by G. C. Fisher and I. Mitchell from 1995). Information on the case of Philip P. comes from court documents from the state of Tennessee.

For details of the Sally Clark case, the best sources are John Batt's book *Stolen Innocence* (Ebury Press, 2005) and Sally Clark's website, www.sallyclark.org.uk. Contemporary newspaper accounts reported her tragic death (see, for example, *The Times*, November 8, 2007). Angela Cannings has given a remarkable account of her own terrible experiences and her brush with Roy Meadow in her autobiography, *Against All Odds* (Little, Brown, 2006).

The Clark case is discussed in almost every book and scholarly article concerned with the misuse of probability in medicine. See, for example, "Conviction by mathematical error? Doctors and lawyers should get probability theory right," *British Medical Journal* 320, no. 7226 (January 1, 2000), or the book *Bad Science* by Ben Goldacre. A fascinating televised lecture by Oxford statistician Peter Donnelly can be viewed at http://www.ted.com /speakers/peter_donnelly.html.

Chapter 2: The Case of Janet Collins

Our primary source for the Collins case was the California state supreme court judgment *People v. Collins*, 68 Cal.2d 319. This fascinating text contains the basic facts of the case, many quotes from original testimony, and the mathematical analysis of the errors made in the original trial. We also studied contemporary newspaper accounts, in particular, articles from the *Los Angeles Times* and the *Independent* shortly after the trial; the article "Trials: The Laws of Probability" in the January 8, 1965, issue of *Time* magazine; and articles from the *Independent* following the appeal judgment in 1968.

A valuable secondary source was the article "Green Felt Jungle," by George Fisher, in the collection *Evidence Stories,* edited by Richard Lempert (Foundation Press, 2006). Fisher actually interviewed both prosecutor Ray Sinetar and mathematician–expert witness Daniel Martinez by telephone in 2005. We also dug into a set of course notes on evidence, law, and reason by

Professor Bruce Hay of Harvard Law School, dating from the spring semester of 2009, on the topic of "Reasonable Doubt."

Chapter 3: The Case of Joe Sneed

It is difficult to find a reference for the details of the Joe Sneed murder trial. Apart from scattered articles in the newspaper archives, this case survives essentially in its very frequent citations in other legal judgments examining mathematical questions raised in court, and of course in the scholarly works on the question. In order to delve more deeply into the case, we contacted the Dona Ana County district court in Las Cruces, New Mexico, where Sneed was tried. For a fee, the court sent us a complete set of photocopies of the archived trial documents, which are the source of nearly everything we were able to learn about the case, with the exception of Kathy Storey's testimony, reported in the *El Paso Herald* on the day after the appeal trial. The court documents are not transcripts—none remain, and the stenographer who took notes is now deceased—but some two or three hundred pages of various affidavits, motions, requests, records of jury selection, letters between lawyers and judges, occasional quoted testimony, and of course the final judgments. From these documents we were able to piece together much information and write a first version of our chapter.

At this point, we realized that we could actually make direct contact with the main mathematical witness in the murder trial, Dr. Edward Thorp. A long telephone conversation with him gave us precious insights into the Sneed case, which allowed us to improve our chapter greatly. We are very grateful to him for all the information he provided to us, as well as a photo of himself from that time.

Chapter 4: The Case of Meredith Kercher

The main source for information about the murder of Meredith Kercher and the facts surrounding this murder and the subsequent arrest and trial of Amanda Knox, Raffaele Sollecito, and Rudy Guede come from original court documents from the two trials in Perugia, Italy. The 427-page "motivations report" submitted after the original verdict by Giancarlo Massei, the judge in the first trial of Amanda and Raffaele, contains an enormous amount of factual detail. Other aspects emerge from the appeal briefs, reports of court sessions, and from the motivations report of Rudy Guede's supreme court appeal trial. Finally, the motivations report submitted following Amanda's and Raffaele's acquittal on appeal by Judge Claudio Pratillo Hellman not only was a source of information, but also contains the particular mathematical example analyzed in this chapter. English translations of most of these documents can be found online at the Perugia Murder File message board (perugiamurderfile.org).

Numerous books have been written about the case, among which we cite Barbie Latza Nadeau's *Angel Face: The True Story of Student Killer Amanda Knox* (Beast Books, 2010), John Follain's *A Death in Italy: The Definitive Account of the Amanda Knox Case* (Hodder & Stoughton, 2011), and of course John Kercher's recent book about his daughter, *Meredith: Our Daughter's Murder and the Heartbreaking Quest for the Truth* (Hodder & Stoughton, 2012), as well as Raffaele Sollecito's first-person account, *Honor Bound: My Journey to Hell and Back with Amanda Knox* (Gallery Books, 2012).

Online blogs and message boards devoted to this unusual and tragic murder proliferate. Perugia Murder File hosts both an ongoing discussion in which facts are analyzed as they emerge and a set of translations of documents from the court file, including court testimony as well as many original writings and statements from the accused. Other sites devoted to the case are True Justice for Meredith Kercher (truejustice.org) and a slew of sites supporting Amanda's innocence (Injustice in Perugia [injusticeinperugia.org], Perugia Shock [perugiashock.com], Friends of Amanda[friendsofamanda.org], and more).

Chapter 5: The Case of Diana Sylvester

Our main source of information for this case, including some of the mathematics, comes directly from the appellant's and respondent's briefs prepared for John Puckett's upcoming appeal.

An in-depth study of the case titled "FBI resists scrutiny of 'matches,'" by Jason Felch and Maura Dolan, was published in the *Los Angeles Times* on July 20, 2008. Another article, "Sex Offender, 74, convicted in 1972 murder," by Jaxon van Derbeken, appeared in the *San Francisco Chronicle* on February 22, 2008. Both are available online, as is the article "DNA's identity crisis,"

by Chris Smith, in *San Francisco Magazine* from September 2008, which tells
the story from Bicka Barlow's point of view, and the article "DNA's Dirty Little
Secret," by Michael Bobelian, in the *Washington Monthly* of March/April
2010. All of these contain interesting information. David Kaye's important article "Rounding Up the Usual Suspects: A Legal and Logical Analysis of DNA
Database Trawling Cases" (*North Carolina Law Review* 87, no. 2 [2009]) gave
important insights into the mathematics of database trawling.

Many articles and blogs have analyzed the mathematics used by both
the prosecution and the defense. Here are some articles concerning the
birthday problem and the Arizona data: Steven Levitt, one of the authors of
Freakonomics, gives his take on the math of the situation in his August 19,
2008, article "Are the FBI's Probabilities About DNA Matches Crazy?" at
http://www.freakonomics.com/2008/08/19/are-the-fbis-probabilities-about
-dna-matches-crazy. An amusing online dispute between National Public
Radio's "Math Guy," Keith Devlin, and Charles Brenner can be found at Arizona DNA Database Matches, http://dna-view.com/ArizonaMatch.htm. On
the more delicate question of the probability that the match found in the
California database is actually the criminal, see the article "Rehash and
Mishmash in the *Washington Monthly*," by David Kaye, at the Double Helix
Law   blog,   http://www.personal.psu.edu/dhk3/blogs/DoubleHelixLaw
/2010/02/rehash-and-mishmash-in-the-washington-monthly.html (February
27, 2010); the article "Guilt by the Numbers," by Edward Humes, in *The
California Lawyer,* available at http://www.callawyer.com/clstory.cfm?eid
=900572&ref=updates (April 2009); and a Quomodocumque blog post titled "Prosecutor's Fallacy—Now with Less Fallaciousness!" by Jordan Ellenberg, simplifying Kaye's explanation, at http://quomodocumque.word
press.com/2010/05/18/prosecutors-fallacy-now-with-less-fallaciousness
(May 18, 2010). Our analysis contains elements of all these, but we do not
fully agree with them, as explained in the chapter.

Chapter 6: The Berkeley Sex Bias Case

The essential source of information for the statistics and mathematical
analysis of the 1973 Berkeley admissions lawsuit was the report by the investigating committee, "Sex Bias in Graduate Admissions: Data from Berke-

ley," by P. J. Bickel, E. A. Hammel, and J. W. O'Connell, published in *Science* 187, no. 4175 (February 7, 1975).

For the particular case of Professor Jenny Harrison, sources included conversations with Berkeley department members (including Professor Harrison, who also provided us with a photo) and press reports of her lawsuit as it progressed. A particularly detailed account was given in the article "Fighting for Tenure: The Jenny Harrison Case Opens Pandora's Box of Issues About Tenure, Discrimination, and the Law," by Allyn Jackson, published in *Notices of the American Mathematical Society* 41, no. 3 (March 1994). Paul Selvin also covered the case and some of the interesting (and rather depressing) effects of its aftermath in two articles in *Science*: "Jenny Harrison Finally Gets Tenure in Math at Berkeley" (July 16, 1993) and "Harrison Case: No Calm After Storm" (October 15, 1993).

Chapter 7: The Case of Lucia de Berk

The website devoted to Lucia's case, http://www.luciadeb.nl/english (for the English-language page), contains a mine of articles and a list of the most important references on the case. Although Ton Derksen's book *Lucia de B.: Reconstruction of a Miscarriage of Justice* (Veen Magazines, 2006) has not been published in English, the website contains a chapter-by-chapter synopsis, a twenty-four-page summary, and the complete translation of chapter 3, concerning the death of baby Amber. Dutch readers can also consult the book *Es werd mij verteld, over Lucia de B.*, by Metta de Noo (Aspekt Ed.) and Lucia de Berk's own book about her terrible experience, *Lucia de B.: Levenslang en tbs* (Arbeiderspers).

This is one of the cases whose mathematical aspect has been most deeply investigated by professionals. It is discussed in chapter 5 of Derksen's book. Piet Groeneboom's blog, http://www.pietg.wordpress.com, contains a prescient entry that dates back to May 2007 titled "Lucia de Berk and the Amateur Statisticians." Richard Gill, a professional statistician who was very active in having Lucia's conviction overturned, has links on his Leiden University homepage, http://www.leidenuniv.nl/~gill, to his own informative comments, slides of technical lectures he has given, and actual research articles that he authored on the subject. Some of these date back as far as

2007, when the errors in court provoked by the testimony of expert witness Henk Elffers began to emerge. An informative although fairly mathematically superficial article called "Lucia de Berk: a martyr to stupidity," by Ben Goldacre, appeared in the *Guardian* on April 10, 2010. The comments to the online version contain a letter from expert witness Henk Elffers to Goldacre and a letter he wrote to the *Guardian,* as well as lengthy and highly relevant comments from Richard Gill. The latter also provided us, via e-mail, discussions with important information and interesting documents, particularly the two original statistical memos by Elffers. Finally, personal communication with Metta de Noo gave us special insight into the workings of the movement that eventually led to Lucia's release; we also thank her for providing us with two photos of Lucia, one taken in her own garden.

Chapter 8: The Case of Charles Ponzi

Charles Ponzi's adventures have been wonderfully documented, first in his 1935 autobiography, *The Rise of Mr. Ponzi,* long out of print but reprinted by Inkwell Publishers (2001), and then in a number of other biographies, of which we appreciated *Ponzi's Scheme: The True Story of a Financial Legend,* by Mitchell Zuckoff (Random House, 2005), which adds a number of new facts to the autobiography. There were far too many interesting and amusing contemporary press accounts of Ponzi's doings to include in the chapter, but the interviews in the local papers, the announcements of his bankruptcy, and his obituary in *Time* were revealing of the popular attitude toward Ponzi during his life. On the story of Bernie Madoff, which should never have happened if the lessons of Ponzi had really been learned, Harry Markopolos' book *No One Would Listen: A True Financial Thriller* (Wiley, 2010) is filled with remarkable, almost incredible information. Like a reincarnation of Ponzi's magnetic personality, Madoff's inexplicable charisma deafened people to what should have been obvious.

Chapter 9: The Case of Hetty Green

There is a great deal of biographical information about Hetty Green on the Internet and innumerable articles about her in the newspaper archives; she was a highly visible figure in her time. An excellent and extremely informative biography is *Hetty: The Genius and Madness of America's First Female Tycoon,* by Charles Slack (HarperPerennial, 2005). Also Daniel Alef's book, *Hetty Green: Witch of Wall Street* (Titans of Fortune Publishing, 2009), contains interesting information. William Emery's *The Howland Heirs* (E. Anthony & Sons, 1919), a book devoted to the genealogy of the Howland family, quotes the most important passages of Sylvia's will verbatim and gives a brief account of the trial and its outcome.

On the subject of Benjamin Peirce and his mathematical analysis of the signatures on Sylvia Howland's will, most sources accept his conclusion more or less unquestioningly. However, a fascinating and much more critical analysis was given by statisticians Paul Meier and Sandy Zabell in their article "Benjamin Peirce and the Howland Will" (*Journal of the American Statistical Association* 75, no. 371 [September 1980]). Our arguments for and against Peirce's use of the binomial model owe a great deal to this unusual paper.

Chapter 10: The Dreyfus Affair

The sources we used for the chapter on the Dreyfus affair are essentially in French. For the actual set of historical events around the Dreyfus case, the definitive book is *L'Affaire Dreyfus,* by Jean-Denis Bredin. For firsthand accounts there are the letters Dreyfus wrote from Devil's Island to his wife and brother during his incarceration (*Lettres d'un innocent* [P. V. Stock, 1898; rprt. Nabu Press, 2010]) and the personal memoirs of Dreyfus' brother Mathieu (*L'Affaire telle que je l'ai vécue* [Grasset 1978]). The full transcripts of the Zola trial are available in print (*Le Procès Zola: Compte-rendu sténographique* [P. V. Stock, 1898]). Newspaper archives contain a mine of contemporary articles on all aspects of the case, frequently accompanied by line drawings.

On the mathematical aspects of the case, sources were rarer and more difficult to locate. One fascinating discovery was a pamphlet published in 1904, authored by A. Bertillon and his assistant Capitaine Valério, containing a complete exposition of the mathematical analysis he applied to the famous bordereau (*Le Bordereau* [Imprimerie Hardy & Bernard, 1904]). The website Poincaré and Dreyfus, http://www.maths.ed.ac.uk/~aar/dreyfus.htm, contains a collection of materials concerning the mathematics involved in the Dreyfus affair, in particular a facsimile of the original 1904 report on Bertillon's work written by Darboux, Appel, and Poincaré; a retyped version in French; a translation into English; and a list of scholarly articles. One interesting introductory article is "Revisiting Dreyfus: A more complete account of a trial by mathematics," by D. H. Kaye (*Minnesota Law Review* 91, no. 3 [2007]). More scholarly articles precisely concerning the Poincaré report can be found in French journals devoted to the history of mathematics; for example "Un mathématicien dans l'affaire Dreyfus: Henri Poincaré," in the Seminar on History of Mathematics of the Institut Henri Poincaré, February 13, 2002, and "Introduction au rapport de Poincaré pour le procès en cassation de Dreyfus en 1904," by Roger Mansuy and Laurent Mazliak, *Electronic Journ@l for History of Probability and Statistics* 1, no. 1 (2005).

CREDITS

All photos not credited below are in the public domain.

Math Error Number 1: The Case of Sally Clark: Motherhood Under Attack
 Page 4, Steve and Sally Clark, © *The Telegraph*
 Page 6, Dr. Roy Meadow, © Getty Images

Math Error Number 3: The Case of Joe Sneed: Absent from the Phone Book
 Page 49, Edward Thorp, courtesy of Edward O. Thorp

Math Error Number 4: The Case of Meredith Kercher: The Test that Wasn't Done
 Page 63, Meredith Kercher, © *Daily Mail/Solo Syndication*
 Page 64, Raffaele Sollecito, © Getty Images
 Page 67, Amanda Knox, © Getty Images
 Page 65, "house of horrors," courtesy of Leila Schneps

Math Error Number 5: The Case of Diana Sylvester: Cold Hit Analysis
 Page 89, Diana Sylvestor, pencil drawing, courtesy of Coralie Colmez

Math Error Number 6: The Berkeley Sex Bias Case: Discrimination Detection
 Page 110, University of California at Berkeley, courtesy of Ken Ribet
 Page 113, Jenny Harrison, courtesy of Jenny Harrison, © Harrison Pugh

Math Error Number 7: The Case of Lucia de Berk: Carer or Killer?

Page 125, Lucia in garden, courtesy of Metta de Noo

Page 133, pencil sketch of Lucia based on newspaper original, courtesy of Coralie Colmez

Page 144, Lucia at ceremony, courtesy of Metta de Noo, © Wijnand Sta

INDEX